Principles for Determining the AIR FORCE ACTIVE/RESERVE MIX

Albert A. Robbert • William A. Williams • Cynthia R. Cook

Prepared for the United States Air Force
Project AIR FORCE • RAND

PREFACE

The mix of active and reserve forces constituting the total Air Force has shifted during the last decade's force drawdown. However, reductions across the active and reserve components of the total force have not been proportional. Moreover, decisions affecting one component may have been made independently of decisions affecting other components. Recent force reductions and transfers among components may not have been guided by a clear and accepted set of principles for determining an end-state force mix.

The shape and size of the force mix can have important implications for the cost, effectiveness, sustainability, and popular and political support of military forces. However, force planners and programmers may not be fully aware of these implications and may therefore make or advocate force-structuring decisions that do not optimally support national interests. This report assembles, examines, and rationalizes a set of principles to help force planners and programmers recognize these implications.

For the most part, fundamental principles applicable to contemporary active/reserve force questions have been hammered out in past policy deliberations. Our contribution was to assemble the principles in a coherent framework and to elicit a review and critique of the framework by current stakeholders and commentators. Toward that end, our research included a forum, held in May 1998, of leaders and thinkers from a number of military, civilian, academic, legislative, and interest-group organizations.

This research was undertaken for the Director of Strategic Planning, Headquarters United States Air Force. It responded to, and benefited from, interest on the part of the Air National Guard and Air Force Reserve advisors to the Director.

PROJECT AIR FORCE

Project AIR FORCE, a division of RAND, is the Air Force Federally Funded Research and Development Center (FFRDC) for studies and analysis. It provides the Air Force with independent analysis of policy alternatives affecting the deployment, employment, combat readiness, and support of current and future air and space forces. Research is performed in four programs: Aerospace Force Development; Manpower, Personnel, and Training; Resource Management; and Strategy and Doctrine.

CONTENTS

FIGURES

TABLES

SUMMARY

INTRODUCTION

What should Air Force decisionmakers consider when making force-mix deliberations across each element of the total force—active, Air Force Reserve (AFR), and the Air National Guard (ANG)? Generally, rational deliberations of force-mix decisions have focused on three factors—cost, military effectiveness, and availability. However, these three factors may not be the only considerations that should apply in determining an appropriate force mix. There is also the issue—understood but often intangible—of how reserve forces help to meet certain social and political objectives important to the Air Force and Department of Defense (DoD) and how the reserve component (RC) captures valuable experience and expertise that would otherwise be lost. In addition, it is necessary to understand why the flow of human capital from active to reserve forces must be kept within feasible bounds. Finally, it is important to understand cost considerations in a disaggregated way; in other words, does the type of mission the Air Force performs favor one component over the other?

This report answers the following two questions: (1) what principles should be considered in force-structure decisions that affect the active/reserve mix? and (2) how do these principles interact with one another? We addressed the first question primarily by reviewing previous studies or commentaries on force-mix issues and found that the relevant principles are generally recognized but have not been assembled into a coherent framework. In addressing the second question, we noted that the principles generally do not prescribe a specific active/reserve mix. Rather, they tend to suggest constraints—the proportions of the mix should be above or below some specified boundary, which may vary as a function of total force size or other factors. When considered simultaneously, these constraints may define a feasible region within which a range of force-mix possibilities would be acceptable. If no such feasible region exists, force-structure planners must choose or compromise between conflicting constraints.

In proposing and discussing these principles, we have observed that active component (AC), AFR, and ANG representatives, and their advocates outside of the Air Force, tend to see the issues through different lenses. We have attempted to avoid a component-specific viewpoint, basing our proposed principles on an overall objective of maximizing the Air Force's contribution to national defense. In some cases, that amounts to optimizing the distribution of available resources within the Air Force's total force. In other cases, it involves enhancing the Air Force's posture for claiming resources (funds, manpower, policy license, political support, etc.) from the larger society. In the latter cases, the worthy objective, we believe, is not to maximize Air Force resources through political manipulation but rather to maximize the quality of the Air Force's linkages to the larger society, relying on the democratic process to govern the resource outcomes.

DESIGNING A FORCE-MIX FRAMEWORK

Based on our review of previous studies and commentaries on the active/reserve mix, we found that six major factors should be included in designing a force-mix framework: (1) social considerations, (2) political considerations, (3) readiness, (4) availability, (5) personnel flow, and (6) cost.

Figure S.1 provides a scheme for integrating these six factors. Arrows on the boundary lines indicate the expected direction of the constraint. The figure suggests that some of these constraints might vary as a function of total force size.

The figure depicts social and political considerations establishing lower-bound constraints. Political utilities depend in part on the total force maintaining a visible presence, with either active or reserve

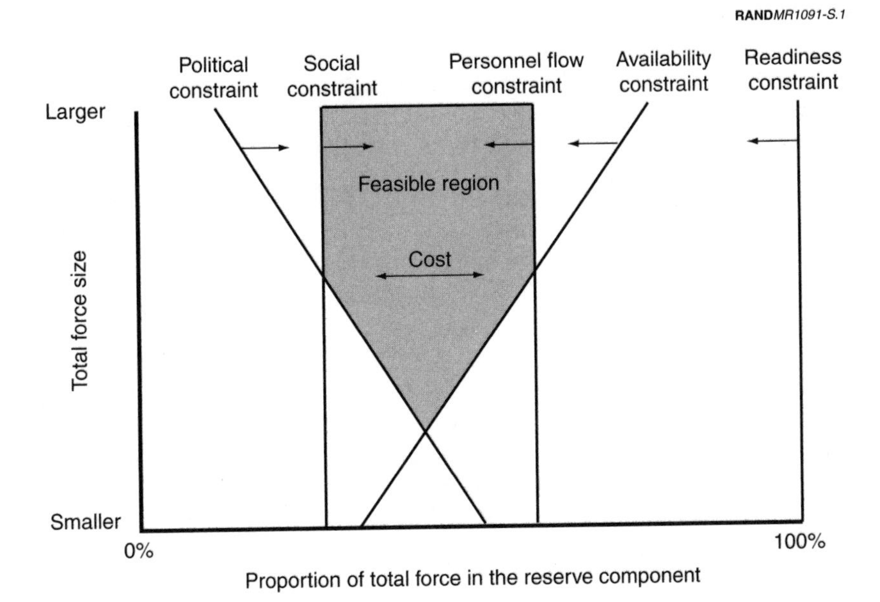

Figure S.1—A Framework for Considering Multiple Force-Mix Principles Simultaneously

forces, in local communities. As total force size decreases, reserve forces will be increasingly called on to provide the presence and must therefore constitute, at a minimum, a larger proportion of the total force. Thus, the political constraint is represented by a sloping line. We suggest that social utilities depend on the reserve forces—a minority within the total force—occupying a large enough proportion of the force to continue to influence the values and culture of the active force majority.

Readiness and availability considerations set upper bounds on reserve forces. We assume a constant demand for rapidly deployable forces that must be met predominantly with active forces. As the total force becomes smaller, this constant demand calls for an increasing proportion of the force to be supplied by the active component (represented by a sloping availability constraint line in the figure). Our hypothesis is that there are generally no appreciable readiness differences between Air Force active and reserve forces. Thus, readiness does not constrain the mix.

The personnel flow constraint also sets an upper bound. The RC depends on a critical flow of human capital from the AC. As the RC proportion increases, it becomes so large relative to the AC that this critical flow can no longer be sustained. In the steady state, this proportion would not vary with total force size.

Taken together, these constraints form a feasible region in which a force mix would simultaneously satisfy the principles associated with the constraints. A specific force mix can then be targeted on the basis of cost considerations. Reserve forces are conventionally viewed as less costly than active forces, indicating that the proportion of the force in the RC should be along the border formed by the upper-bound constraints. However, reserve forces may be more costly than active forces in meeting operations other than war (OOTW) and smaller-scale conflict (SSC) demands, so that the direction in which cost considerations drive the force depends on the need being met. If OOTW/SSC demands cannot be met with the least costly major theater war (MTW) force structure, it will be necessary for force structure planners to choose between a larger, more reserve-intensive force that better meets MTW demands, and an equal-cost, smaller, more active-intensive force that better meets ongoing high-tempo demands.

BUILDING THE FRAMEWORK—WHAT THE ANALYSIS TELLS US

In examining the available evidence, we found that the loci of some of these constraints are currently unknowable and that others are mission- or component-dependent. Where possible, we derived notional results using available data or what we believe to be reasonable estimates thereof, focusing for purposes of illustration on fighter force structure.

In our analysis of how social and political considerations constrain the force mix, we introduce terminology that allows us to more precisely label the associated constraints. We find that a social *identification, embeddedness, and investment* (IE&I) constraint is sloped, so that as the total force decreases in size and is more geographically concentrated, the RC will play an increasing role in maintaining contact with the larger society. However, the precise location of this lower boundary cannot be determined. We also find that a *minority status* constraint—driven by research on the boundary between a token and minority level of representation for a separately identifiable subgroup within an institution—sets a floor for the RC at about 20 percent of the total force.

Based on our analysis of readiness and availability, we find (as hypothesized) a readiness parity between the AC and RC, which leads to no readiness constraint. Availability, however, is limited for the RC because of its predominantly part-time workforce. Because of greater limits on cumulative deployment time and duration of deployment relative to the AC, the RC is less available for meeting contingency requirements. Thus, the availability constraint, like the social IE&I constraint, is sloped (although in the opposite direction). In other words, as the total force declines in size, assuming force employment demands remain constant, a decreasing proportion of the total force can be placed in the RC.

Based on our modeling of personnel flows between the active and reserve forces, we find that we can establish a rough upper bound on the proportion of the RC in the total force, assuming other related parameters are known. These parameters vary by mission and mission design series (MDS) and also by differences between ANG and AFR in their perceived ability to absorb inexperienced undergraduate pilot training (UPT) graduates.

Finally, in terms of cost, our analysis argues for a larger proportion of the total force in the RC when contemplating MTW scenarios and a smaller proportion when contemplating SSC and OOTW scenarios. Decisionmakers must weigh the tradeoffs between meeting MTW and SSC/OOTW needs.

APPLYING THE FRAMEWORK TO AN ANG CASE—A NOTIONAL EXAMPLE

Figure S.2 depicts the framework for the ANG case applied to the fighter force structure, where notional personnel flow constraints might allow the RC to occupy up to 42 percent of the total fighter force. This creates a feasible region to the right of the social constraint. The feasible region might be reduced if an availability constraint came into play or if a decisionmaker were to supply a judg-

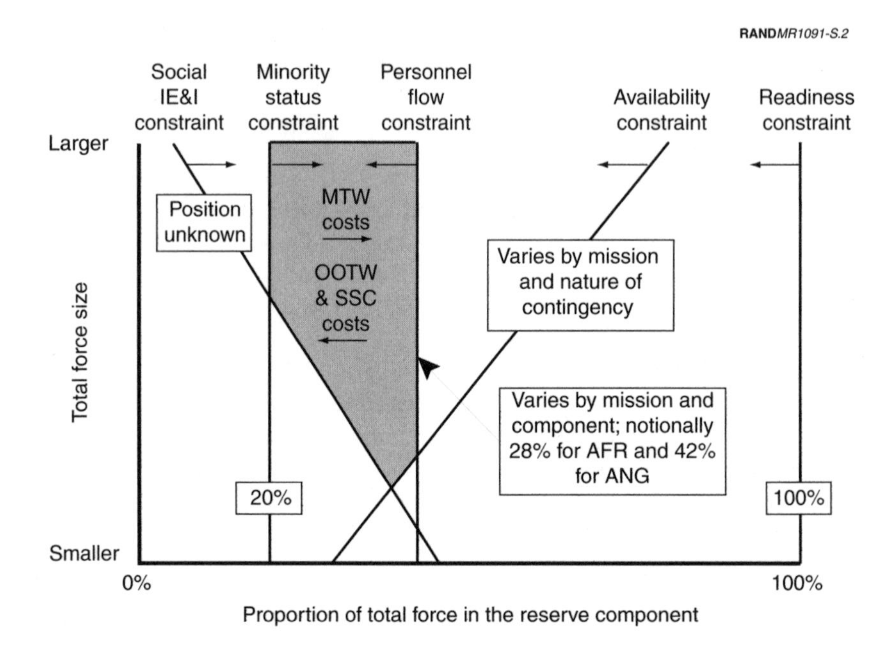

Figure S.2—Notional Values for Force-Mix Constraints: An ANG Case

mental locus for the political constraint. Within this feasible region, cost-conscious decisionmakers would gravitate toward a 42 percent mix if they were primarily concerned about preparedness for MTW scenarios or toward a 20 percent mix if they were concerned with meeting current contingency deployment needs. It is possible, of course, to weigh cost more heavily than either the personnel flow or social constraints. In that case, decisionmakers might drive the mix above 42 percent, consciously accepting a degradation in experience levels and readiness. Alternatively, they could drive the mix below the 20 percent RC minority status constraint, possibly compromising RC members' capacity to influence the values and perceptions of AC members.

If personnel flow or availability were evaluated using different parameters, those constraints could conceivably lie to the left of the RC minority status or social IE&I constraints. There would be no feasible region. In such a case, decisionmakers would have to compromise between conflicting objectives. The most likely outcome would be to deemphasize the minority status and social IE&I constraints because the case for applying the former in the present context is less compelling and the locus for the latter is unknown.

We again stress that the specific force-mix results reported here are notional. Where possible, we used input values that we judged to be approximately correct, recognizing that we did not have the resources in this project to obtain or derive analytically rigorous inputs, especially when the inputs are likely to vary across missions. Also, because of variations across missions or MDSs, force-mix decisions cannot be made in the aggregate. They must be made for each mission or MDS individually.

CONCLUSIONS

Although the results reported here are notional, we believe our research provides two concrete contributions to the force-mix decision process. The first contribution is a framework for integrating the range of considerations that decisionmakers face and gaining perspective on the arguments offered by various interest groups hoping to influence the force mix. The second contribution is a roadmap for more detailed research into specific mission/MDS force mixes or a general model that incorporates mission/MDS-specific inputs.

Perhaps our most significant finding is that cost considerations cut in opposite directions depending on whether the force is being optimized for major theater war preparedness or for peacetime contingency operations. In our view, peacetime contingency demands must be given more weight in force-mix decisions, especially in MDSs experiencing high deployment-related stress.

ACKNOWLEDGMENTS

Inspiration for this project came from Brig Gen John Harvey, USAFR (ret.), and Brig Gen Joseph Simeone, ANG, who served as Air Force Reserve and Air National Guard advisors to the Air Force Director of Strategic Plans during a critical period in the project. Lt Gen David W. McIlvoy, Air Force Director of Strategic Planning at the inception of the project; his successors, Maj Gens Charles F. Wald and Norton A. Schwartz; and their deputy director, Dr. Clark Murdock, supported the project. Lt Col Cal Hutto, our Air Staff point of contact, provided essential coordination and support. Maj Carl D. Rehberg helped us with Air Force Reserve issues, while Maj Greg Riddlemoser and Mr. Gary Taylor provided similar support regarding the Air National Guard.

Forty individuals joined us at RAND in May 1998 for a forum that explored the issues addressed in this report. We thank all of the participants in that forum for helping us to focus on the right issues and to place them in appropriate perspectives. In particular, we thank Mr. Brian Sharratt, Deputy Assistant Secretary of the Air Force for Reserve Affairs; Maj Gen (ret.) Roger Sandler, president of the Reserve Officer Association; Maj Gen Sam Carpenter, military executive for the Reserve Forces Policy Board; Mike Higgins, a professional staffer for the House National Security Committee; and Larry Korb, a senior fellow at the Brookings Institute. Other participants in the forum, to whom we also owe our appreciation, include several state assistant adjutant generals for air, senior mobilization augmentees, senior staff members from the Air Staff and several major commands, academicians, and fellow members of the RAND staff.

Within RAND, Bob Roll, our program director at the inception of the project, provided strong leadership and valuable insight in shaping the project. Colleagues Don Palmer and Carl Dahlman provided helpful input and advice. Paul Steinberg and Jeanne Heller contributed immeasurably to the readability of the report. Reviews by Jack Graser and Dick Buddin sharpened our thinking in several areas.

Any remaining errors are, of course, our own.

ACRONYMS

AC	active component
AFR	Air Force Reserve
ANG	Air National Guard
DOC	designed operational capability
DoD	Department of Defense
FH	flying hours
FWE	fighter wing equivalent
IE&I	identification, embeddedness, and investment
MDS	major design series
MTW	major theater war
NPS	nonprior service
OOTW	operations other than war
OPTEMPO	operations tempo
ORI	operational readiness inspection
PAA	primary aircraft authorized
PERSTEMPO	personnel tempo
POL	petroleum, oil, and lubricants
PS	prior service
QDR	Quadrennial Defense Review
RC	reserve component
RPI	rated position identifier
SAAM	special assignment airlift mission
SORTS	Status of Resources and Training System
SSC	smaller-scale conflict

TDY	temporary duty
UPT	undergraduate pilot training
UTC	unit type code

Chapter One

INTRODUCTION

BACKGROUND

In Air Force and Department of Defense (DoD) force-structure decisionmaking, each element of the total force—active, Air Force Reserve (AFR), and the Air National Guard (ANG)—has proponents that often seek to maximize the resources devoted to it. Proponents include senior leadership within the components, Congress, and stakeholders external to DoD, such as associations that advocate the interests of the various components. In this oftentimes competitive environment, resource allocation decisions would better support public interests if they were guided by a set of principles for maximizing a total Air Force contribution to national defense.

Moreover, the force mix has changed significantly during the last decade. Figure 1.1 shows that active component (AC) strength reductions were proportionally greater than reserve component (RC) strength reductions during the last decade (which has shifted the mix toward a greater proportion in the RC). In fiscal year (FY) 1988, the two RCs—the ANG and the AFR—together constituted 25 percent of total Air Force strength and 11 percent of total operating costs. In FY 1998, the RC constituted 33 percent of the total strength and 16 percent of the total cost. In our review of the literature and in our interviews with AC and RC decisionmakers, we found no evidence that this shift occurred as part of a conscious force-mix strategy. Rather, it occurred as a result of many decisions, taken independently, regarding active component (AC) and reserve component (RC) force structure. We argue that there are good reasons for mak-

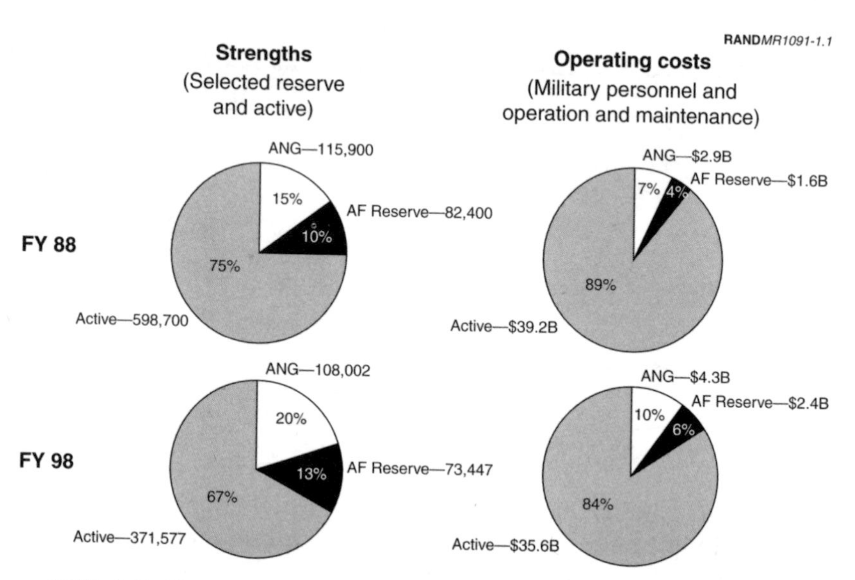

SOURCES: FY 88 and 89 National Defense Authorization Act (Public Law 100-180), FY 88 DoD Appropriations Act (Public Law 100-202), FY 98 National Defense Authorization Act (Public Law 105-85), and FY 98 DoD Appropriations Act (Public Law 105-56).

NOTE: The sum of military personnel appropriations and operations and maintenance appropriations is used as a proxy for operating cost. This provides only a rough indication of operating costs, because the military personnel appropriation includes the cost of military personnel engaged in nonoperating activities, such as procurement. Other appropriations—Procurement; Research, Development, Test, and Evaluation; Revolving and Management Funds; Military Construction; and Family Housing—are either unrelated or less clearly related to the operating costs of the components.

Figure 1.1—Strengths and Operating Costs

ing force-structure decisions affecting the force mix in an integrated rather than an independent way.

Generally, rational deliberations of the force mix have focused on three factors—cost, military effectiveness, and availability. As the data in Figure 1.1 suggest, and as will be demonstrated in more detail later in the report, force structure costs less in the RC than in the AC. Additionally, Air Force RC units, in contrast to some reserve forces in other services, generally meet or exceed AC levels of military effectiveness. However, reserve forces are less *available* than active forces—except in small numbers, they cannot be deployed as rapidly as active forces; there are strict statutory limits on how and for how long they may be called up involuntarily for active duty; and there are practical limits on how long and how often they can be employed voluntarily.

These considerations can be combined to form an implicit principle for determining a cost-effective force mix—maximize the RC proportion, subject to satisfying availability demands that generally require active forces.

However, unit operating costs, military effectiveness, and availability are not the only considerations that apply in determining an appropriate force mix. There is also the issue—understood but often intangible—of how reserve forces help to meet certain social and political objectives important to the Air Force and DoD and how the RC captures valuable experience and expertise that would otherwise be lost. In addition, it is necessary to understand why the flow of human capital from active to reserve forces must be kept within feasible bounds. Finally, it is important to understand cost considerations in a disaggregated way; in other words, does the type of mission being performed favor one component over the other?

OBJECTIVES AND APPROACH

In preparing this report, we sought to answer the following questions:

- What principles should be considered in force-structure decisions that affect the active/reserve mix?
- How do these principles interact with one another?

In reviewing previous studies or commentaries on force-mix issues, we found that the relevant principles are generally recognized but have not necessarily been assembled into a coherent framework. We also found that some principles have been articulated for military forces in general, and thus need to be tailored to the Air Force case.

In addressing the second question, we noted that the principles generally do not prescribe a specific active/reserve mix. Rather, they tend to suggest constraints—the proportions of the mix should be above or below some specified boundary, which may vary as a function of total force size or other factor. When considered simultaneously, these constraints may define a feasible region within which a range of force-mix possibilities would be acceptable.

PERSPECTIVE

In proposing and discussing these principles, we have observed that AC, AFR, and ANG representatives, and their advocates outside of the Air Force, tend to see the issues through different lenses. We have attempted to avoid a component-specific viewpoint, basing our proposed principles on an overall objective of maximizing the Air Force's contribution to national defense. For most of the principles we advocate, that amounts to optimizing the distribution of available resources within the Air Force's total force.

When examining social and political issues, we find that the operant objective is to enhance the Air Force's posture for claiming resources (funds, manpower, policy license, political support, etc.) from the larger society. The possibility exists that success in this endeavor could come at the expense of the other services, of domestic programs, or of other social interests. Evaluating these welfare economic implications is beyond the scope of our study. Consequently, the perspective we adopt in evaluating social and political issues is not an economic one. Rather, we take an organizational ecology perspective—military organizations will not obtain needed resources if they fail to cultivate appropriate linkages to the larger society. The worthy objective, we believe, is not to maximize Air Force resources through political manipulation but to maximize the quality of the Air Force's linkages to the larger society, relying on the democratic process to govern the resource outcomes.

SCOPE

The end product in this research is *not* a specific force mix. As we shall demonstrate, the appropriate force mix is contingent on a number of variable factors. Our objective is to identify the relevant principles, leaving to decisionmakers and their staffs the task of applying the principles in specific force-structuring actions.

ORGANIZATION OF THE REPORT

Chapter Two describes in general terms the broad factors we have found to be important in force-mix decisions and provides a model to enable the factors to be considered simultaneously. In Chapters Three through Six, we discuss the implications of these factors in greater detail, showing—based on our analysis—where the locus of each factor or constraint is in the model. Chapter Seven gives our conclusions and recommendations.

Chapter Two

FORCE-MIX PRINCIPLES—AN OVERVIEW

Based on our review of previous studies and commentaries on the active/reserve mix, we found that the relevant issues can be aggregated into six major factors. In this chapter, we briefly describe those factors and indicate in general terms how they influence the ideal force mix. We also present our framework for considering the factors simultaneously. Subsequent chapters examine the individual factors more thoroughly.

SIX FACTORS IN MAKING ACTIVE/RESERVE MIX DETERMINATIONS

The six factors we identified are (1) social considerations, (2) political considerations, (3) readiness, (4) availability, (5) personnel flow, and (6) cost. Since social and political considerations are closely related, we discuss them below together; the same holds true for readiness and availability, which are also discussed together.

Social and Political Considerations

Reserve forces may have social and political utilities that differ from or complement active force utilities. To realize these utilities, we suggest that the proportion of the total Air Force supplied by the RC must exceed some critical mass.

These utilities are related to a number of considerations. There is a strong historical and constitutional presumption that at least part of the nation's military forces should be provided by citizen-soldier militias rather than regular forces. Although rooted in political concerns prevalent in the nation's revolutionary period, reliance on militia forces may have contemporary benefits.

- Reservists are more fully integrated into the larger society than active-duty members, enabling them through personal contact to extend public awareness and trust of military institutions.
- Call-up of reservists for real-world employment is subject to stronger political checks and balances than the employment of active-duty forces, thus discouraging military involvement that lacks public support and broadening support for employments that are undertaken.
- Reserve units, especially those of the ANG, are less geographically concentrated than active units. (One might expect that the greater geographic representation of the RC would make it more demographically representative than the AC, but our analysis in Chapter Three did not find it markedly so.) Representativeness in any form contributes to public trust in government institutions.
- By increasing the number of veterans in the society, reserve forces increase the proportion of key public policy decisionmakers and influencers who have military experience and are thus more likely to take informed positions about military issues.
- ANG units provide an efficient and effective source of disciplined manpower to satisfy state missions (disaster relief, civil disturbance, etc.) .

Readiness and Availability

Because reserve forces rely heavily on part-time participants who have full-time occupations, they are less available than active-duty forces. They may also be less ready than active-duty forces because of limited opportunities for training, particularly unit-sized training exercises that cannot be effectively compressed or segmented into weekend drill periods. In practice, these limitations apply more to Army, Navy, and Marine Corps reserve forces than to Air Force reservists, many of whom can effectively train in small aggregations (individuals or crews) and who can also be readily integrated into active forces without involuntary unit call-ups. Nonetheless, active forces must remain large enough to meet rapid-deployment needs and to provide sustained involvement in operations that exceed statutory limitations on the duration of call-ups.

Active forces also face some readiness-limiting conditions. Turnover in active-duty flying squadrons is higher than in reserve squadrons because of rotations into and out of cockpit duties and to permit greater absorption of new pilots entering the rated force. RC pilots often have previous AC experience. As a result, AC pilots have on average less weapons system experience than their RC counterparts.

Personnel Flow Considerations

To meet their manpower needs, reserve forces rely heavily on a flow of trained and acculturated personnel from active-duty forces. It is unlikely that reserve forces could find sufficient qualified nonprior service (NPS) recruits to meet all their needs, given that nonprior service recruits generally must agree to an initial period of active duty for training lengthy enough to complete recruit and initial skill training. If the maximum acceptable nonprior service input to reserve forces can be determined and if active-force separation and reserve affiliation rates are known, an upper limit on the ratio of reserve to active forces can be determined.

Cost

Active and reserve force operating costs have often been compared on a per-unit or per-aircraft basis. That is, the operating costs of reserve units are compared with those of similar active units. This cost comparison approach assumes equal availability and employability of active and reserve units. Such an assumption is appropriate for employment of forces in major theater war (MTW) or some large-scale military operations other than war (OOTW) scenarios, where call-up of reserve forces can make them fully substitutable for active forces for many requirements (those that are compatible with statutory call-up limitations).

However, in most OOTW or smaller-scale contingency (SSC) scenarios, this cost comparison approach is not appropriate.

Sustained deployments, such as those related to peacekeeping in Bosnia or enforcement of no-fly zones in Iraq, generate demands that, given acceptable durations and frequencies of deployment, are different from MTW demands. Short of call-up, reserve forces face a limit on acceptable frequency and duration of employment that is much tighter than that of active forces. Thus, in meeting these demands, reserve forces are not fully substitutable for active forces. An appropriate cost comparison for OOTW and SSC scenarios must be based on outputs useful in these scenarios. A relevant output is the number of days per year that an aircraft with appropriate aircrews and logistic support can be deployed. The cost per output is then computed as annual aircraft operating costs divided by deployable aircraft days per year.

INTEGRATING THE SIX FACTORS

Figure 2.1 provides a scheme for integrating the six factors discussed above. The figure depicts a set of constraints on the proportion of the total force that is in the RC. Arrows on the boundary lines indicate the expected direction of the constraint. The figure suggests that some of these constraints might vary as a function of total force size. In subsequent chapters, we will indicate what we believe to be the approximate loci of these constraints in some contexts.

The figure depicts social and political considerations establishing lower-bound constraints on the proportion of the total force in the RC. Political utilities depend in part on the total force maintaining a visible presence, with either active or reserve forces, in local communities. As total force size decreases, reserve forces will be called upon to provide the presence, and they must therefore constitute, at minimum, a larger proportion of the total force. Thus, the political constraint is represented by a sloping line. The social constraint suggests that the reserve forces must occupy some minimum constant proportion of the force in order to influence the values and culture of the total force.

Readiness and availability considerations set upper bounds on reserve forces. We assume a constant demand for immediately and continuously available forces that must be met primarily by using active forces. As the total force becomes smaller, this constant de-

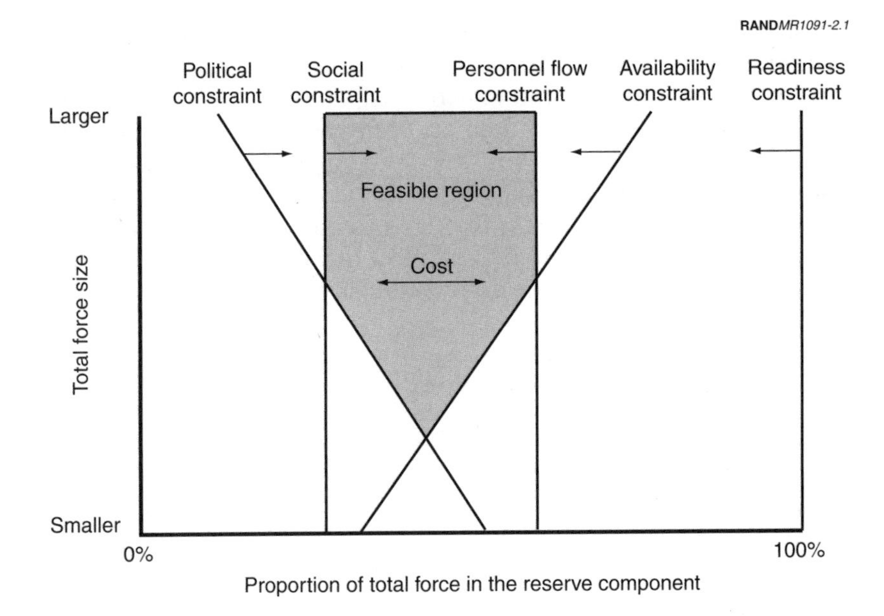

Figure 2.1—A Framework for Considering Multiple Force-Mix Principles Simultaneously

mand calls for an increasing proportion of the force to be supplied by the active component (represented by a sloping availability constraint line in Figure 2.1). As discussed above and in more detail later, there are generally no appreciable readiness differences between Air Force active and reserve forces. Thus, readiness does not constrain the mix.

The personnel flow constraint also sets an upper bound. The RC depends on a critical flow of human capital from the AC. There is some ratio of AC to RC size beyond which this critical flow can no longer be sustained. In the steady state, this ratio would not vary with total force size.

Taken together, these constraints form a feasible region in which a force mix would simultaneously satisfy the principles represented by the constraints. A specific force mix in the region can then be targeted on the basis of cost considerations. As mentioned above, reserve forces have been conventionally viewed as less costly than active forces, indicating that the force mix should be along the border formed by the upper-bound constraints. However, if reserve forces are found to be more costly than active forces in meeting OOTW and SSC demands (see Chapter Six), the direction in which cost considerations drive the force will depend on the need being met. It may be necessary for force-structure planners to choose between a larger, more reserve-intensive force that better meets MTW demands, and an equal-cost, smaller, more active-intensive force that better meets OOTW/SSC demands.

Note that we depict no feasible region at smaller force sizes. In this force size range, the goal of keeping reserve forces large enough to obtain political utilities might conflict with the goal of keeping enough active forces to meet rapid deployment demands. Similarly, it is possible that in some contexts the upper-bound personnel flow constraint will lie to the left of the lower-bound social constraint, so that there is no feasible region at any force size. When there is no feasible region, force-structure decisionmakers must make tradeoffs between conflicting principles.

Chapter Three

SOCIAL AND POLITICAL CONSIDERATIONS

INTRODUCTION

The armed forces of the United States do not and cannot operate in isolation from the larger American society. They are part of a polity whose needs they must serve. Moreover, to win the trust and support of the larger polity, the armed services must be widely perceived as serving important social needs. Without this trust and support, the armed forces will not get the resources they need to function effectively.

We argue that the Air Force can secure support by understanding and managing three kinds of linkages between the military and society (Kestnbaum, 1998). These linkages take the form of a shared value system, a shared social structure, and shared interests and attachments. We further argue that because the RC has considerably more opportunity for interaction with civilian society, it is better situated than the AC to develop these three linkages. The RC can communicate the desires and expectations of civilian society to the total force and can, in turn, communicate the missions and needs of the armed forces to civilian society, thus limiting overall isolation of the armed forces from society.

In this chapter, we first explain the theoretical framework of how each of these linkages works to increase the attachment between the armed forces and society. We then examine a number of social and political factors that have a bearing on the strength of these linkages. Finally, we assess how these social and political considerations should operate to constrain the force mix.

THEORETICAL FRAMEWORK

The first potential linkage between the armed forces and society is a shared value system. Shared values in a democratic society might relate to such issues as diversity and inclusion, citizenship rights in relation to obligations, closeness of the military to the people it is supposed to protect, responsiveness of the government to its people, the protection of democracy from centralism and tyranny, and limitations on adventurism. The operative mechanism to establish this linkage is *identification*, whereby people can look across institutional divides and find others who share their values or ideals. Citizens who identify with their armed forces are more likely to support them.

The second potential linkage is a network of shared social structures through which military members are intertwined with their civilian counterparts. Shared social structures can be found in the workplace, schools, churches, community service or political organizations, or even through being in common social positions, such as middle-class taxpayers in a small town. Compared with the AC, members of the RC are likely to have many more such shared structures with the civilian communities of which they are a part. The operative mechanism to establish this linkage is *embeddedness*, whereby people who are separated by institutional boundaries in one sphere (military versus civilian) are linked with common institutions in other spheres. An armed force embedded into larger society will have more opportunities to understand and be understood by civilians.

Finally, the RC promotes greater shared interests and attachments between those in the armed forces and civilian society, the third linkage. For example, veterans maintain an interest in the military; the number of veterans in society is arguably greater with a larger RC. Also, the mobilization of reservists generates interest among others in their communities. The operative mechanism to establish this linkage is *investment*, through which people become interested in persons and institutions by virtue of their connections and attachments to these people. A citizenry invested in its armed forces is more likely to support them.

SOCIAL AND POLITICAL FACTORS

A number of social and political factors have been offered as arguments for reserve forces:

- The maintenance of state militias as entities separate from a national armed force is written into the Constitution.
- The RC increases the public's awareness and trust of military institutions.
- Public reaction to reserve call-ups provides a check on excessive use of the military.
- The RC is more representative of society than the AC.
- The RC increases political support for the armed forces.
- The National Guard serves specific state roles.

Each of these social and political factors would work through the mechanisms of identification, embeddedness, and/or investment to increase the connections between society and the military.

Beyond the fact that such factors argue for the existence of the reserves, it is also reasonable to consider such factors in force-mix decisions, supplying decisionmakers with compelling reasons to maintain some minimum proportion of the force in the RC. It may be more difficult to quantify how these social and political considerations should affect the force mix than it would be to quantify how other factors such as cost, effectiveness, or personnel flow should do so. Yet without the linkages and support that these social and political considerations embody, the military will be less able to gather the resources it needs. Perhaps more important, the considerations offer something of a basis for maintaining an RC within a democratic society.

Militia-Nation Considerations

The tradition of citizen soldiers in the United States dates back to before the nation was born, and then further back into Anglo-Saxon tradition in England. Part of our romantic understanding of the Revolutionary War is that of farmers laying down their plows and picking up their muskets to drill and then to serve. The Constitution of the United States reflects this tradition and clearly lays the groundwork for the existence of part-time soldiers, as the excerpts in Table 3.1 demonstrate.

Section 8 of Article I gives Congress the power to federalize the militia as a means to achieve broader government objectives such as security and stability. However, the militia is clearly not a federal force—since there are certain rights and responsibilities reserved to the states, such as officer appointments and training. In fact, Section 2 of Article II distinctly separates the regular forces, which now in-

Table 3.1

The Militia in the Constitution and the Bill of Rights

Article I, Section 8	The Congress shall have Power To lay and collect Taxes, Duties, Imposts and Excises, to pay the Debts and provide for the common Defense and general Welfare of the United States; but all Duties, Imposts and Excises shall be uniform throughout the United States; To provide for calling forth the Militia to execute the Laws of the Union, suppress Insurrections and repel Invasions; To provide for organizing, arming, and disciplining the Militia, and for governing such Part of them as may be employed in the Service of the United States, reserving to the States respectively, the Appointment of the Officers, and the Authority of training the Militia according to the discipline prescribed by Congress; And To make all Laws which shall be necessary and proper for carrying into Execution the foregoing Powers, and all other Powers vested by this Constitution in the Government of the United States, or in any Department or Officer thereof.
Article II, Section 2	The President shall be Commander in Chief of the Army and Navy of the United States, and of the Militia of the several States, when called into the actual Service of the United States.
Amendment II	A well regulated Militia, being necessary to the security of a free State, the right of the people to keep and bear Arms, shall not be infringed.

clude the Air Force as well as the Army and Navy, from the militia, or National Guard.

This constitutional language was a compromise between two factions of the document's drafters—the Federalists, who wanted a strong national government, and the Anti-Federalists, who wanted to ensure states' rights. The Federalists inserted the clause that grants Congress the power to call forth the militia. The Anti-Federalists wanted to make sure the states would have access to the militia to balance the powerful central government, so the Second Amendment was incorporated into the Bill of Rights. The language in this amendment is often understood in a limited sense as the right to bear arms. However, the right to bear arms should be understood in the context of maintaining a local militia that can be used by the states.

The militia was understood by drafters of the Constitution to be a crucial means to certain ends that cannot be better served by enlarging the standing army because the two institutions are fundamentally different. Kohn (1997) suggests that the militia is an armed countervailing power to prevent the regular military from becoming too strong and to ensure that civilians have control over military affairs. He sees this countervailing power as necessary to prevent tyranny of a strong central government based on military control and the adventurism of such a government attempting to increase its span of control by trying to conquer other nations. The founding fathers saw examples of these risks all over Europe.

Another end that the militia serves is one of citizenship, as suggested by Kestnbaum (1997). Democratic society is strengthened when rights of participation in the democracy are earned in the defense of the nation. Also, a democratic armed force must remain rooted in the people by making sure that a substantial portion do not see themselves strictly as career military but instead identify with civilians and plan to return to civilian life.

Though the Constitution calls for a militia, it does not offer direct, specific purchase on the question of sizing the National Guard. At its heart, the Constitution is a document embodying the shared values of our society, and the existence of the RC is an expression of these values. The function of the RC in this instance is to help maintain the democracy and enhance its value. Decisionmakers need to be aware of this function when sizing the total force.

Public Awareness/Trust of Military Institutions

The RC serves to promote public awareness and trust of military institutions by providing civilians contact with the military and by providing military members contact with civilian society.

Civilian Contact with the Military. The contact hypothesis suggests that one role reservists play is communicating the culture, structures, and goals of the military to a wider public. This role has not gone unrecognized within the RC community. McDonald (1996) exhorts reserve officers to develop links with the community and to try to enhance the image of the reserves through the media. "Young officers represent an important link between the armed forces and the civilian society and are first-class military ambassadors. They are key players in promoting a broader understanding of the importance of our military defense" (p. 34).

Contact between the military and society provides an entrance for positive military values into the broader culture. For example, military sociologist Charles Moskos and his associates (e.g., Moskos and Butler, 1996) have long made the argument that the military has a higher percentage of African American managers than any other employment sector of the U.S. economy. These managers are, of course, the members of the officer corps. In the military, whites are much more likely to report to, and take direct orders from, blacks than they are in the civilian economy. RC members who are managed by minority officers in the military will be able to recognize and communicate the value of diversity in their civilian jobs.

Civilian contact with the military is enhanced by the fact that the RC, particularly the ANG, is by design far more geographically dispersed than the AC. As indicated in Figure 3.1, 75 percent of the Air Force AC is concentrated in 13 states, whereas 75 percent of the RC is spread over 25 states. For operational reasons related to heavy deployment demand, the AC can be made more efficient and less stressed by concentrating it on a smaller number of larger-scale installations. If such rebasing were to occur within the AC, the RC's relatively greater geographical dispersal would take on even greater

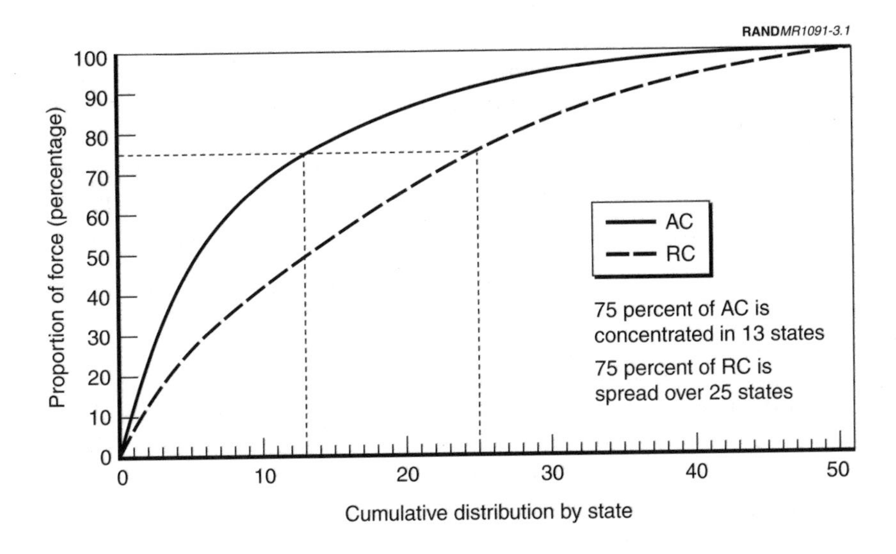

Figure 3.1—Reserve Components Are Geographically Dispersed

importance as an avenue to increase opportunities for contact between civilians and the military.

Military Contact with Civilians. The flip side of public support for military institutions is military members' understanding of the larger society. Although active-duty members of the military may have significant contact with civilians in their daily lives and jobs, they may also be isolated from them, especially if stationed on a remote base or abroad. The military branches have taken considerable care to develop cultures that reflect certain values, with an eye toward making a better, stronger, and more cohesive fighting force. These values may not be shared or, if shared, followed by larger society. Ricks (1997) tells of Marines after boot camp being faced with a kind of culture shock when they go home on leave. Civilians are "a bunch of freaks" (p. 233); "overweight, and a little sloppy" (p. 228); "self-destructive, not trying, just goofing around" (p. 229); "losers" (p. 229); "people with obnoxious attitudes, no politeness whatsoever, nasty" (p. 231). One Marine, fresh out of boot camp says "Defending my country? Well, it's not really my country. I may live in America, but the United States is so screwed up" (p. 236).

The quotations, although not a representative sample, are telling and are a cause for concern. Members of the military are sworn to protect and serve the larger society. The potential danger lies in an armed force that decides that it is above society or that society is not worthy of protection. While we do not consider this a likely outcome, it is the worst-case consequence of a military that becomes too remote and disconnected from the general citizenry and their values.

In fact, members of the military are very different from civilians in terms of their political affiliation. In a 1997 Olin Institute paper (also cited in the *Wall Street Journal*, 1997), Holsti found that the military is notably conservative and partisan. In 1976, 33 percent of the military and 25 percent of the civilian opinion leaders surveyed identified themselves as Republicans.[1] By 1996, 67 percent of the military opinion leaders were Republicans, whereas only 34 percent of the civilian leaders were. This significant, radical shift to the right is even more pronounced among younger military opinion leaders: 92 percent of those born after 1954 are Republican. The *Wall Street Journal* article cites an unnamed three-star general who claims that the "single greatest danger facing the U.S. military today [is] the possibility that a politicized military will stay that way, growing less and less responsive to civilian control over time."

Reservists offer a bridge between the military and larger society because, as full-time employees within and generally longer-term residents of their respective communities, they enjoy greater embeddedness in shared social structures than their AC counterparts. Whereas AC airmen and officers may also have contact with civilians, their contacts are generally less extensive and less well developed than that of their RC counterparts. The RC is better situated than the AC to make known the Air Force's missions and needs to civilians

---

[1] In this study, a survey sample was drawn from 4000 opinion leaders whose names had been derived from such general sources as *Who's Who in America* and *Who's Who of American Women*, as well as more specialized directories listing leaders in occupations that are underrepresented in *Who's Who*, including media leaders, politicians, military officers, labor leaders, State Department and Foreign Service Officers, and foreign policy experts outside government. The military sample included students at the National War College and a smaller number of senior uniformed Pentagon officers whose names were drawn randomly from the *Congressional Directory*.

and in a better position to understand the values and interests of civilians and convey these back to the total force.

Implications for the Force Mix. The force-mix implications of increased opportunity for military/civilian contact are related to how the views and ideas of a minority group are respected by the majority. In this case, the issue is whether a better appreciation of the values and interests of the larger society found among members of the RC (a minority) can be effectively communicated to members of the AC (a majority). Insight can be found in the literature on organizations. Kanter (1977) offers a typology of minority groups based on their level of representation in an organization. Her work focuses on women in the workplace, but the reasoning can be extended to any situation where less-represented individuals are trying to make an impact in a larger group.

In Kanter's system, a *uniform* group is one where members are all in one category. A *skewed* group is one where most of the people are of one type, perhaps making up 85 percent of the whole. Members of the minority group would be rare enough to appear as *tokens* and would face heightened performance pressures, since their successes may be discounted but their failures highly publicized and scrutinized. More seriously, social isolation would make it "difficult for [these members] to generate an alliance that can become powerful in the group" (p. 209). In *tilted* groups, the split is less severe, with perhaps 65 percent of members in one group and 35 percent in the other. Kanter characterizes the larger group in this range as a *majority* and members of the smaller group as a *minority* rather than as tokens. Here, "minority members have potential allies among each other, can form coalitions, and can affect the culture of the group. They begin to become individuals differentiated from each other as well as a type differentiated from the majority" (p. 209). At ratios of 60:40 through 50:50, the majority and minority members are balanced and can become distinctive "subgroups that may or may not generate actual type-based identifications" (p. 209).

For the RC to be taken seriously—to be able to form a coalition that can be heard—this analysis would suggest that maintaining the status of minority rather than token is necessary. Kanter is not clear on the proportion at which members of a subgroup cease to be tokens and achieve minority status.[2] Indeed, this can be expected to vary by situation. However, one possible reading of the work is that a subgroup acquires minority status with as little as 20 percent or as much as 40 percent of the population. This indicates that the RC should constitute a minimum proportion of the total Air Force in the 20–40 percent ranges to ensure it has sufficient "voice." This assumes that the ANG and AFR can join together in a coalition to represent the viewpoints of the citizen-soldier.

Kanter's arguments about tokenism and how proportions of minorities affect their perceived influence were based on observing females in the workplace in the early 1970s. In her example, company management did not provide the leadership required to reduce discrimination against women within the organization. In many cases, women found it difficult to do their jobs effectively because of efforts to subvert them at all levels of the organization. The level of minority representation she offers as a point at which minorities can be heard (35 percent) is probably a function of this problematic situation.

In contrast, RC members performing their jobs among AC members might not be as noticeable as females would be in a predominately male environment. Indeed, it is questionable whether AC officers and airmen in fact view RC individuals as being of distinct and lower status. Also, the Air Force does have experience integrating a minority group into its ranks. The history of removing racial barriers blocking African American participation in the armed services offers a fascinating example of how strong leadership enabled an organization to give lie to the argument that social cohesion resulting from racial similarity is a critical factor in unit performance (Rostker and Harris, 1993). Strong civilian and military leadership that focused on legislating acceptable behaviors of whites rather than on changing attitudes helped create the integrated Air Force that we observe today.

---

[2]Other research on gender tokenism finds a different range of effects. South et al. (1982) find that "token women are not found to face more severe organizational pressures than nontokens" (p. 587). Yoder (1991) finds that the studied pressures on women "occur only for token women in gender-inappropriate occupations," whereas in the Air Force, RC members are not in inappropriate occupations. Izraeli (1983), however, generally supports Kanter's work on tokenism.

It is extremely doubtful that members of the RC are in a position analogous to that of African Americans before integration (or of women in Kanter's study). In contrast, the RC is well integrated into the functioning of the total force. Moreover, the message remains that a strong, effective, and educated leadership can ensure that members of the RC are treated with respect. Thus, we believe that a constraint toward the minimum (20 percent) of Kanter's "minority status" range of thresholds would be sufficient to ensure that the RC has effective voice in the total force.

Linking Force Employment to Public Support

After the Vietnam War, the military developed its *total force policy*, which it has maintained to the present day. Binkin (1993, pp. 110–111) offers a considered discussion of the basis for this policy and its viability.

As related by Binkin, General Creighton Abrams, after the armed forces' Vietnam experience, advocated a close operational association between the active Army and the RC to keep the AC from being sent to a war without the involvement of the RC. The RC would bridge the gap between the active military and American citizens, so that the active military would be less isolated in case of war. Hence, "if reserves must be activated in order to sustain active forces in anything more than limited contingencies, presidents will be less inclined (and politically less able) to become involved in military actions without extensive national debate and political consensus." (Lacy, 1986; also cited in Binkin.)

The total force policy increases the possibility that civilians will be acquainted with someone who is serving in the theater of war, and possibly someone who becomes wounded or killed. In short, since members of the RC are embedded in society, their friends and coworkers will have a higher probability of being directly tied to someone making a sacrifice for the country. The total force policy brings the war home to a larger number of civilians. If only military career professionals were involved, the boundaries around who gets killed or wounded could keep the war on an intellectual and less-emotional level for those civilians without friends or coworkers who are serving and sacrificing. Thus, using the RC ensures an involved society.

Binkin (pp. 149–151) tests this proposition using data from the Persian Gulf conflict, reaching much the same conclusions as a RAND report (RAND, 1992, pp. 95–97). (In fact the two authorities cite each other on this topic.) Binkin plots Gallup poll data on public support for the war against reserve mobilizations, showing that support for the war declined as reserves were being called up. He does not definitively state that there was a cause-and-effect relationship between the two factors, but he suggests that support might have declined further if the conflict had been longer and the number of American casualties had increased. Public opinion did not necessarily act as a brake in the short and relatively unbloody (for Americans) conflict. RAND cites anecdotal evidence that "mobilization of reserves also mobilized support of the war" and that employers supported their reservist employees (p. 96). However, there is no strong evidence either way that integrating the reserves with the active military helps maintain support or diminishes it.

RAND (1992) concludes that if decisionmakers consider that integration of the components of the force is important and necessary, then this is a *political* reason to shape the force in such a way. The resulting interaction produces a citizenry that knows and cares more about the military and the institution and may be more likely to participate knowledgeably in any public debate about force employment. People may be more likely to support funding for the armed forces so that those they know will be better prepared in the event of military action.

However, these analyses offer little or no purchase on the proportion of the total force that must be in the RC. If the thesis underlying the total force policy is valid—that using the RC imposes a check on inappropriate military action and creates public support for those actions that are undertaken—researchers have offered no hypotheses on how many people must have contact with RC members to obtain those effects.

Representative Force Issues

Krislov (1974) suggests that one method of securing broad social support for government policy and action is to draw a representative segment of society into the government. Doing so promotes both investment in the values of and identification with the interests of the government on the part of all segments of society. Applied to the Air Force force-mix question, this perspective would argue to increase proportions of the force that are more demographically representative and decrease those that are not. To determine the relative representativeness of the various components, we reviewed DoD demographic data on officers and enlisted personnel both at current strengths and among new accessions (DoD, 1997).

Gender Diversity. Figure 3.2 shows the percentage of women in the officer corps as well as new officer accessions in fiscal year 1996. Figure 3.3 does the same for the enlisted ranks.

Of the three branches, the AFR has by far the highest percentage of active female officers, at over 24 percent, while the ANG has the least diverse officer corps of the three components, being only 13.4 percent female.

All three components are attracting female officers at a higher percentage than their current representation, with new female officers

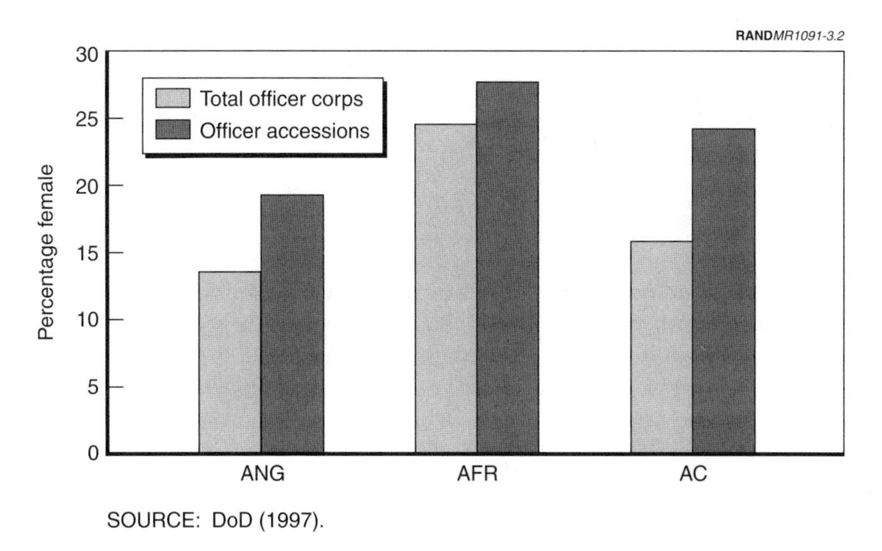

Figure 3.2—Percentage of Female Officers and Officer Accessions in the Total Force Among Three AF Components, FY 1996

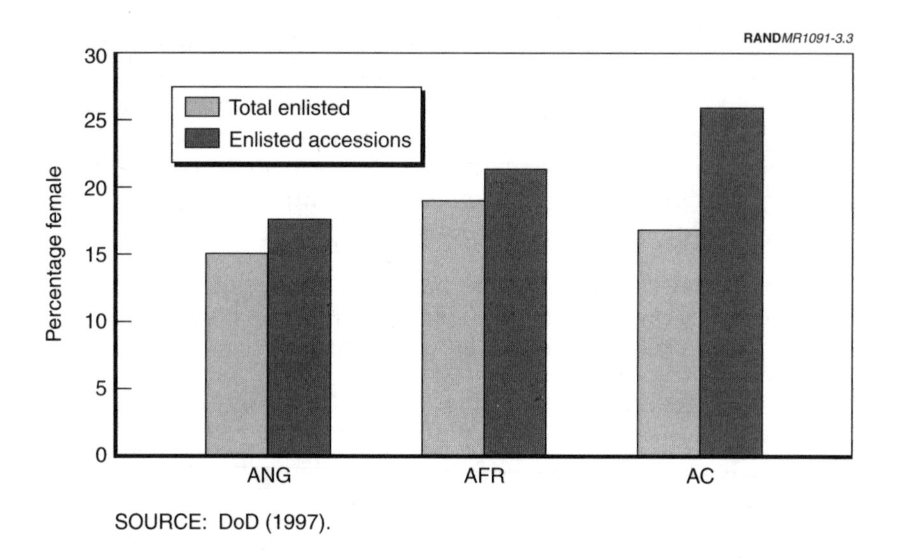

SOURCE: DoD (1997).

Figure 3.3—Percentage of Female Enlisted Members and Enlisted Accessions in the Total Force Among Three AF Components, FY 1996

in the AFR making up almost 28 percent of the total. If the trend is maintained, it will lead to a gradual increase in the percentage of female officers in the three branches. Thus, the level of gender diversity among officers in the Air Force could be increased by making the AFR larger relative to the other two components.

For the enlisted corps, the AFR is again the most gender-representative component. In fiscal year 1996, 19.1 percent of the members of the AFR were female, whereas women made up 16.9 percent of the AC and 15.1 percent of the ANG. The ANG again has the least gender diversity in the Air Force.

However, of the three components, the AC is recruiting the largest percentage of enlisted females, at 26 percent. If this trend continues, the AC should surpass the AFR as the most gender-representative component. The AFR lags behind, with 21.4 percent of its new recruits being female. Again, the ANG attracts the smallest proportion of new female recruits, at 17.7 percent. However, all three components are recruiting females at higher levels, which over time will increase the percentage of female enlisted airmen. Thus, the level of gender diversity among enlisted personnel in the Air Force could be increased by making the AFR and AC larger relative to the ANG.

We note that our analyses of gender representation were conducted at an aggregate rather than an occupational level. Some occupations have been, historically, more female-intensive than others. Thus, it is possible that gender differences among the components reflect differences in the occupational mix among the components.

Racial/Ethnic Diversity. Figure 3.4 breaks down the numbers for the officer corps and for officer accessions of the three components in fiscal year 1996 in terms of racial/ethnic diversity. Figure 3.5 provides the same information for enlisted personnel.

As the figure shows, there are few major racial/ethnic differences among AC and RC officers. The ANG has the highest percentage of Hispanic officers, whereas the AC leads in percentage of blacks. However, the differences are not great.

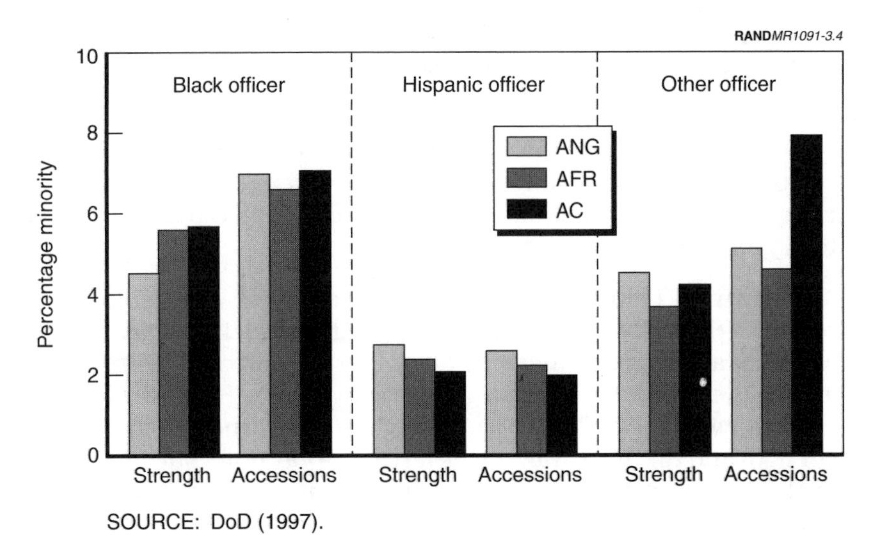

Figure 3.4—Percentage of Minority Officers and Officer Accessions Among Three AF Components in the Total Force, FY 1996

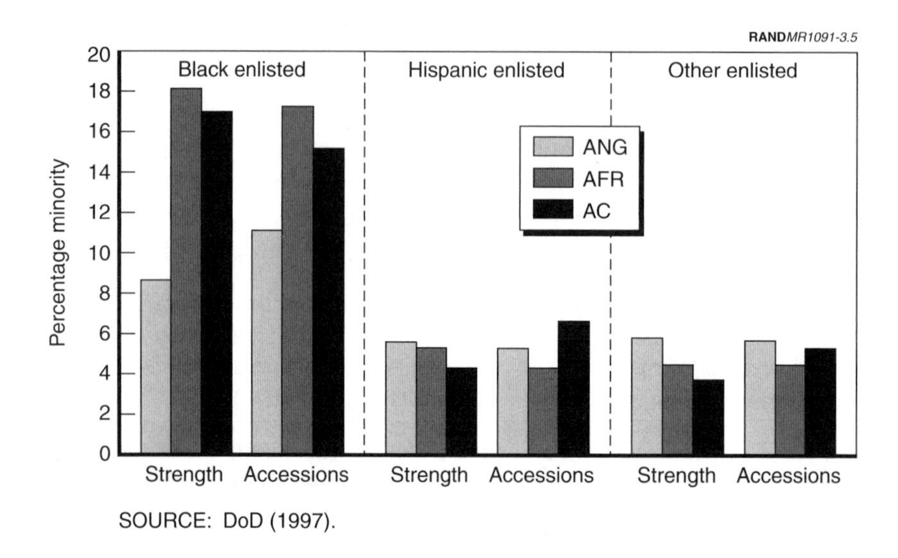

Figure 3.5—Percentage of Minority Enlisted Personnel and Enlisted Accessions Among Three AF Components in the Total Force, FY 1996

For officer accessions, there are again only slight differences in recruitment patterns of minorities. The ANG exceeds the other two components in its recruitment of Hispanics. The AC has the highest total proportion of minority accessions.

As shown in Figure 3.5, the enlisted ranks are markedly more racially/ethnically diverse than the officer corps. In particular, the AFR has a high percentage of blacks, at 18 percent. Differences among the components regarding participation by Hispanics are less marked, with the ANG having the highest proportion in its current strength and the AC recruiting the largest proportion. The ANG leads in the "other" category, which includes Asians, Native Americans, and Pacific Islanders. We cannot draw clear conclusions about which component does the best job of recruiting and retaining minorities.

Implications for the Force Mix. We found that the RC is more diverse than the AC in some gender and racial/ethnic categories and less diverse in others. Thus, representativeness does not argue for shifting the proportion of the force toward either the RC or the AC.

Influence of Veterans in Society

Holding force costs constant, a larger RC proportion results in a larger total force and, arguably, produces more veterans (members with some military service).[3] The influence of these veterans can be positive in defense-related matters. In addition, veterans have ties to others in society and can thus increase understanding of the military among those with whom they come into contact.

Butler and Johnson (1991) studied how Americans felt about fiscal support for the military (spending on arms and foreign aid, in particular), the obligation to serve, the overall quality of the military, and minorities in the military and opportunities for minorities serving. They analyzed data from the General Social Survey, a biennial national survey of adults not living in institutional settings (such as hospitals, prisons, and military barracks). They pooled data from 1982, 1983, and 1984 to generate a sample with a larger number of veterans and African Americans. Generally, they find that, holding other factors constant, veterans, older people, and southerners are more likely to support the military, while more highly educated people are less supportive of the military. Characteristics having little effect include race and income. Of importance here is their finding that military service increases support for the military.

Ivie, Gimbel, and Elder (1991) analyzed data on men and women who were born in the 1920s to see if military experiences in World War II and Korea affected their attitudes toward the military. They find that being a veteran or being married to a veteran, having a child who served in the military, and maintaining social ties with friends from the service increase support for military preparedness.[4]

---

[3]Holding costs constant, a larger RC proportion results in a smaller AC and an RC that increases by more than the decrease in the AC. Some proportion of the larger RC requirement would be met using nonprior service resources. If those nonprior service resources turn over at the same or higher rates than the smaller number of AC resources they displace, the result will be more veterans. Although we do not have separate turnover for prior-service and nonprior service reservists, we note that turnover in the RC is generally higher than turnover in the AC.

[4]Support for military preparedness is measured by a five-point scale ranging from "strongly disagree" to "strongly agree," indexing support for the following four statements: (1) A strong defense should be the number-one priority today, ranking above social needs and a balanced budget; (2) Registration for the draft is needed to ensure a strong America; (3) Military training should receive strong support in our colleges and

Similarly, Butler and Johnson (1991) find that veteran status is positively related to support for the military.

Military experience is increasingly rare among members of Congress. Within the last 25 years, the percentage of members with any military experience has fallen from 70.6 to 35.8 percent.[5] With fewer veterans in Congress, there is a greater possibility that military appropriations will fall short of needs.[6]

Political Influence of the RC

Members of the RC can use their extensive political networks to garner national support for the armed forces. The ANG, in particular, has members in all 50 states who can lobby their congressional representatives in support of their goals. Even without active lobbying by the RC, congressional interest in maintaining a local military presence, perhaps because of jobs, will enhance the likelihood that Congress will vote in support of particular AC or RC goals.

One example is the perpetual overfunding of C130 transport aircraft procurement. Year after year, Congress, perhaps lobbied by either the aircraft manufacturer or local reserve components, funds procurement of more C130s than the Air Force requests.[7] The extras find a home in the RC, where in fact a large portion of the airlift mission exists.

---

high schools; and (4) All men above 18 should be required to take a certain amount of military training.

[5]Data for the 93rd Congress are derived from *Roster of United States Congressional Officeholders and Biographical Characteristics of Members of the United States Congress, 1789–1993: Merged Data File*, 9th Inter-University Consortium for Political and Social Research, 1993, Ann Arbor, Michigan. Data for the 105th Congress are derived from *Congressional Universe* (worldwide web service), Congressional Information Service, Bethesda, Maryland (accessed December 17, 1998).

[6]As stated in the Introduction, the objective of fostering a larger number of veterans in society and in government is not to *maximize* military resources but rather to help create conditions in which democratic social and political processes result in an *appropriate level* of military resources.

[7]This raises the question of whether the Air Force in fact *relies* on knowledge that this will happen when it puts together its budget requests for aircraft acquisition. If programmers know that Congress is going to force a certain number of airlift aircraft on the Air Force, the Air Force can ask for more fighters or bombers than if it had to husband its resources more carefully.

According to this argument, RC political influence should be harnessed in support of all new weapon acquisitions by planning for initial introduction of new weapons in both AC and RC units. But should force-mix decisions be shaped to invoke political support? Our view is that the public interest is generally not well served when an agency attempts to bend national priorities toward its own ends through political advocacy rather than shaping its mission to what the citizenry deems important. This is particularly true if a less effective or efficient force mix were adopted to gain political advantage.

We distinguish such *direct* political RC influence from a more indirect sort, which is the development of public support for the armed forces through voting and other manifestations of political preferences. The military needs public support to sustain itself in an environment of limited resources, where defense is just one of many public goods competing for tax dollars. The RC, with its ties to the larger society, is well positioned to communicate the importance of defense and national security policy to a citizenry whose more immediate concerns and interests may lie in other directions. We argue in this report that the RC's indirect influence—through the processes of identification, embeddedness, and investment—plays a significant role in generating public support for the military.

State Missions of the ANG

Unlike the other components of the armed services, the National Guard has a state role. It can be called upon by state governors to offer emergency assistance in a disaster that may present problems to an overwhelmed citizenry. Examples are snowstorms, floods, earthquakes, and fires, as well as emergencies resulting from social unrest.

Brown, Fedorochko, and Schank (1995) examined the nature of the state missions of the National Guard to determine if the Guard had sufficient manpower to fulfill them. They collected survey data from 49 of the 54 National Guard entities (in all 50 states, plus the District of Columbia, Guam, Puerto Rico, and the U.S. Virgin Islands). In addition, they conducted 15 site visits to study in depth state requirements for the Guard and how well they had been met. The study included both the Army National Guard and the Air National Guard.

The authors examined cases where the Guard was called out to respond to state emergencies, including Hurricane Andrew that struck Florida and Louisiana in 1992, Hurricane Iniki that damaged Hawaii in 1992, the enormous Midwestern floods in 1993 that put huge portions of Wisconsin, Missouri, Iowa, Illinois, South Dakota, Nebraska, Kansas, and North Dakota under water, and the 1992 Los Angeles riots that followed the acquittal of the police officers who had been accused of brutalizing Rodney King. In each of these cases, National Guardsmen numbering in the thousands were called up to respond to the crisis conditions.[8]

The staffing of the National Guard was more than adequate to perform all the state missions for which it was called upon. The Guard usually backed up state resources and the Federal Emergency Management Agency (FEMA) rather than serving as a front-line response to disaster. There may not be enough National Guard capacity to fully resolve peak disasters, but these peaks are relatively rare, and the Guard is one tool in an adequate package of emergency relief that is sufficient to respond to civil emergencies. Brown and his colleagues do not recommend increasing the size of the Guard to respond to peak disasters, which they characterize as an uneconomical approach. Sizing the Guard to deal with rare disasters would mean creating a force that is underused the vast majority of the time. If increased capability should be considered necessary, regional pacts between the state Guard organizations could institutionalize and ease the sharing of resources among the states. The example of the Oklahoma Air Guard airlifting feed to cattle in New Mexico while the New Mexico Guard responded to other aspects of a heavy snowstorm shows the viability of this option.

Brown et al. (1995) report that only a small fraction of ANG units or members serve on state missions in any given year (and in many years there is no requirement for their services). Thus, even a

---

[8]Major disasters tend to require resources more likely to be found in the Army National Guard than in the Air National Guard. However, the ANG contributes special capabilities, such as airlift and civil engineering, in addition to general-purpose manpower.

significantly reduced RC would not negatively affect the ability of the ANG to perform its state missions.

State missions of the ANG garner a great deal of positive publicity from the media and support from the public, increasing interest and investment in the institution. Moreover, citizens who may have little or no contact with the military or its members in usual times may have significant contact during disasters, further deepening this means of attachment between civilian society and the armed forces. This is probably not a sufficient reason to set a floor on the size or proportion of the ANG in the total force.

HOW SOCIAL AND POLITICAL CONSIDERATIONS CONSTRAIN THE FORCE MIX

Our current research seeks to determine the degree to which social and political considerations should constrain the force mix. We illustrate our findings using the two constraint lines depicted in Figure 3.6. Note that the lines in Figure 3.6 are labeled differently from the corresponding lines in Figure 2.2. The labels introduced here more fully reflect the vocabulary and concepts introduced in Chapter Three. One line represents the RC proportion of the total force needed to provide a sufficient level of social identification, embeddedness, and investment (IE&I) linking the armed forces to the larger society. The other line represents the minimum proportion that the RC must occupy to have a meaningful level of representation and influence within the total force.

The IE&I constraint line is sloped but its precise position cannot be determined. We can theorize that as the total force decreases in size and is more geographically concentrated, the RC would play an increasingly important role in maintaining contact with the larger society. The RC would have to occupy an increasingly larger proportion of the total force to provide the required mass in a sufficient number of communities. This explains the slope of the line. However, we have no basis for estimating the mass or community penetration needed to obtain these benefits. Thus, the position of the line is unknown.

A minority status constraint is shown at 20 percent. As discussed earlier, this is at the low end of Kanter's range of thresholds between

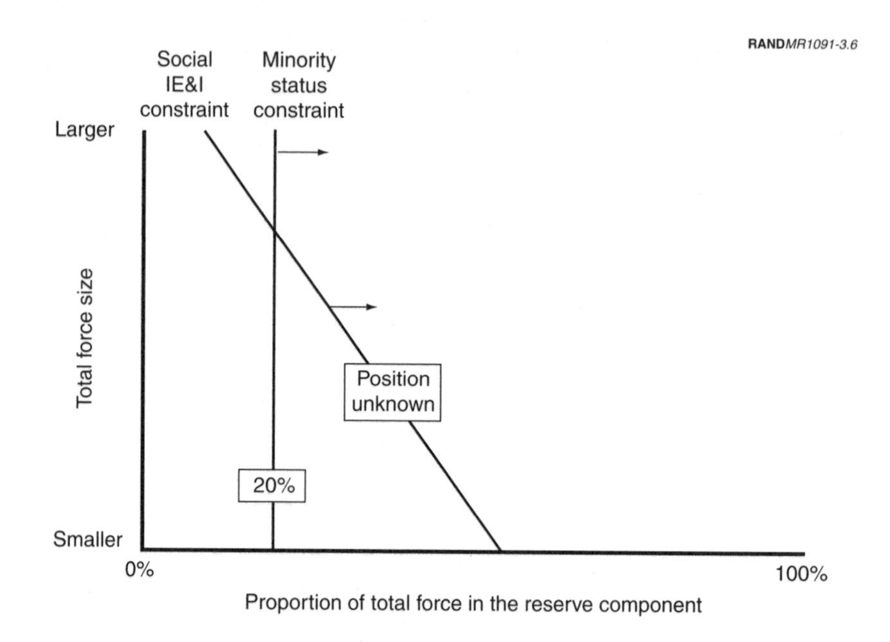

Figure 3.6—Locus of Political and Social Constraints on the Force Mix

a token and minority level of representation for a separately identifiable subgroup within an institution. We offer this constraint with the caveat that Kanter's theory was developed after observing demographic minorities that are more clearly distinct than are the active and reserve components of the armed services, and in a situation where leadership did not work to ensure that the minority was taken seriously. Generalizing the theory to apply it in the force-mix context must be done with caution.

Finally, we offer a caveat that Air Force force-mix implications cannot be considered in isolation from those of the other armed services. In many cases, the social and political functions that are part of the logic of a strong RC are served similarly by the Army and the Air Force National Guards and Reserves. The Army RC is much larger than the Air Force RC, and presumably how the Army addresses the force-mix question would have more of an effect on the feedback loop between the military and society than would the force mix within the Air Force.

Chapter Four

READINESS AND AVAILABILITY

In this chapter, we discuss effectiveness: how well RC units serve military needs relative to similarly configured AC units. To answer this question, we considered two aspects of effectiveness: the *readiness* of a unit to perform its intended military functions and the *availability* of a unit for employment by national command authorities. Readiness depends on the unit's access to resources (personnel and equipment) and to processes (training and maintenance) needed to keep these resources combat-ready. Availability combines elements of responsiveness (how soon is a unit available) and duration (for how long is it available). If AC and RC units systematically differ in either of these two aspects, effectiveness will be affected by the force mix.

READINESS

Generally, modern RC air assets receive much praise for their combat effectiveness. For example, RC units and aircrews served during operations Desert Shield and Desert Storm with little readiness differences between them and their AC counterparts (RAND, 1992, pp. 56–57). Similar evidence exists that RC units perform well when deployed for current peacetime contingency operations.[1]

[1]A current joint force combat operations center director said that there was no discernible difference between RC and AC units deployed in his area of responsibility. He claimed that only minimal local area checkout was required for experienced RC units. Another observer attributed only average or below-average performance to RC combat rescue units, some of which arrived at forward operating areas lacking basic combat rescue skills. This observer said RC rescue units may not have adequate access to

Although actual performance in combat is the best readiness indicator, it cannot be observed during peacetime. Thus, we must look at available peacetime readiness indicators to determine how RC units compare with AC units. Some measures of readiness are available from the joint readiness reporting structure—the Status of Resources and Training System (SORTS). Other indicators include operational readiness inspections (ORIs) and exercises where RC units perform with their active counterparts. SORTS measures inputs to readiness—personnel status, equipment status, and training—whereas inspections and exercises tend to measure outputs or outcomes more akin to actual combat capabilities.

Peacetime Readiness Indicators

As an input measure, SORTS provides a limited means for benchmarking units. Air Force units generally maintain a higher natural state of readiness than the other services in both the AC and RC, and SORTS reveals little difference between RC and AC air units.

ORI results also show little difference between AC and RC units, although the process for RC units can be somewhat different from that for AC units.[2] The inspection team includes augmentees from the gaining command and much effort is invested in making sure that the process mirrors the active process for like units. Few differences between the AC and RC were found in ORI results for the years 1992–1996 (see Figure 4.1). Overall ratings (on a scale of 1 to 11) are slightly lower for RC units, although they are still well in the satisfactory range. For fighter units, the overall scores are somewhat higher for RC units than for AC units.[3]

---

ranges and other forces necessary to train to the same level as active-duty combat rescue units. In addition, these units may be more actively engaged with state and local missions, which build basic airmanship but do little for the more demanding combat rescue mission where a high degree of coordination with supporting forces is required. The combat rescue mission has a relatively high mission operations tempo for the total force.

[2]Doing well on the ORI in most RC units is highly desired, just as it is in the active force. One reservist told RAND that the first question asked during his initial hiring interview was his willingness (and ability, given civilian employment) to make himself available for the unit ORI.

[3]Unpublished data gathered from Air Combat Command, office of the Inspector General, by RAND researchers Willard Naslund and Craig Moore.

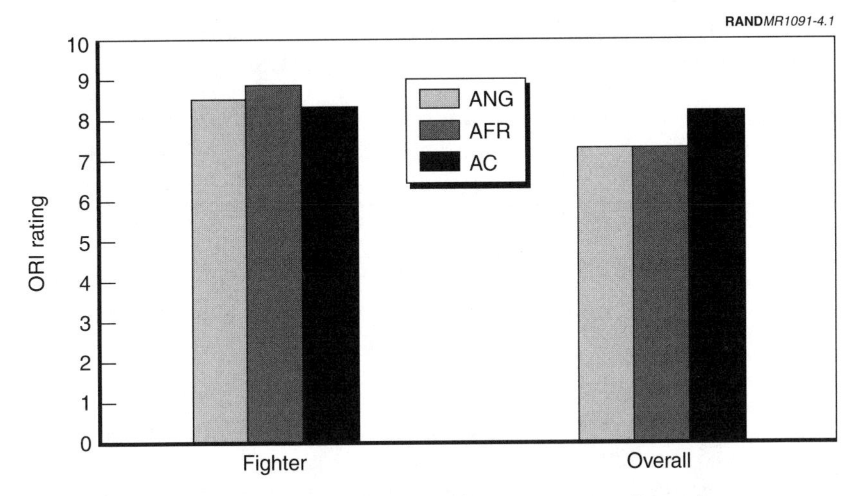

SOURCE: Air Combat Command, Office of the Inspector General.

Figure 4.1—Average ORI Ratings, 1992–1996

RC units have historically done very well in various combat crew competitions. This could be the result of an RC unit having higher average total career and mission design series (MDS) flying time and greater experience overall than a sister AC unit has. But competitions may not be an appropriate readiness gauge for RC or AC performance, because the flight and maintenance crews who participate are the best-of-the-best. However, a higher average number of total career flying hours and more time with the unit are major strengths that RC aircrews bring to the total force. This difference in experience will increase as anticipated shortages of AC pilots emerge.[4]

---

[4]Operations personnel at a mixed-force active wing told RAND that some combat fighter squadrons had 40–50 percent of personnel with less than two years of active flying experience beyond their initial mission qualification (RAND interviews, April–May 1998).

Underlying Factors—Experience, Training, and Operational Integration

We identified three factors that underlie the RC's readiness parity with the AC. First, as mentioned above, RC personnel generally have more unit-specific and aircraft-specific experience than their AC counterparts. Second, RC aircrews, despite having fewer available duty days, come close to achieving qualitatively what AC units achieve with higher numbers in their training programs. Third, relatively small RC elements can be readily integrated with other RC and AC elements to form provisional units for employment.

Experience. Experience levels in RC units are generally significantly higher than they are in AC units for several reasons. First, many members of RC units have prior AC experience. Second, RC aircrews tend to remain in cockpit flying duties far longer than AC aircrews, who must rotate between staff and cockpit assignments. As an example, Table 4.1 shows indicators of experience for pilots in two B52 squadrons, one in the RC and one in the AC.

Training. The training program for an RC unit generally contains the same categories of training missions as the training program of an AC unit with a similar mission, but with fewer missions required in some cases. Table 4.2 shows, for example, the number of annual training missions for comparable RC and AC units. Annual mission requirements are identical to maintain a basic mission-capable status (for pilots in staff positions), but combat mission-ready status (for pilots in line cockpit positions) requires fewer missions in the RC. Some observers believe that generally higher experience levels

Table 4.1

Average Flying Hours and Combat Experience of Pilots in Typical RC and AC B52 Units

Indicator	RC	AC
Total hours	3266	1809
B52 hours	2244	1446
Instructor/evaluator hours	621	464
Proportion of pilots with combat experience	60%	12%

SOURCE: Office of Air Force Reserve, Headquarters USAF, data as of April 1998.

Table 4.2

Ready Aircrew Program (RAP) Annual Training Missions

	Basic Mission Capable		Combat Mission Ready	
	Inexperienced	Experienced[a]	Inexperienced	Experienced[a]
		F16, Block 40		
AC	72	60	116	96
ANG	72	60	90	76
		A10		
AC	72	60	102	90
ANG	72	60	90	72

SOURCE: Directorate of Training, Headquarters Air Combat Command RAP tasking messages for July 1998 to June 1999.

[a]Experienced pilots are those who exceed an established flying-hour threshold that varies by MDS.

permit RC units to maintain the same proficiencies as AC aircrews even with fewer training missions.

Operational Integration. To be considered ready in the prevailing environment of contingency deployments, RC units must be able to function effectively in the package sizes at which they are commonly available. As discussed in more detail below, RC participation in most contingency deployments depends on voluntary participation rather than mobilization. Deployable packages of RC personnel and equipment therefore tend to be of less than full squadron scale.

Fortunately, air operations and their direct support allow smaller packages from multiple units to be assembled to form provisional units at employment sites without unacceptably losing operational effectiveness. An example is the ANG RAINBOW deployment of ANG Block 42 F16C/Ds. Aircraft, manpower, and resources are combined from three units to deploy in-theater in a unified mission package. Each unit commander has flexibility in meeting his deployment commitments while enabling the ANG to deploy a sufficiently sized unit in peacetime.[5]

---

[5]The RAINBOW concept demonstrates that integration of smaller units is possible without major problems. However, unless units are exercised on a regular basis, unit commanders may feel more comfortable with members of their own squadrons and logistics support. Recent RAINBOW deployments have concentrated on deploying

AVAILABILITY

In the past, availability was relatively easy to define for RC units. RC units were written into operational plans in a manner similar to AC units. RC units generally flow into a deployment schedule alongside AC units, assuming timely mobilization of the RC units. After mobilization, RC units become full-time assets for a specified duration. However, this is not a complete picture of how joint force commanders are using forces today. Recurring and long-duration peacetime force employments require the RC to participate in a nonmobilized status. Thus, availability of reservists under both mobilized and nonmobilized conditions must be considered.

Availability When Mobilized

Legal limits on mobilization are contained in Title 10, US Code, Section 12301-12305. The main provisions, which apply to all reserve components collectively (Army, Navy, Air Force, Marine Corps, and Guard), are as follows:

- In time of war or national emergency declared by Congress, reserve units and individuals not assigned to units may be ordered to active duty for the duration of the war or emergency and for six months thereafter.
- At any time, a reserve unit or individual not assigned to a unit may be ordered to active duty for not more than 15 days a year.
- In time of national emergency declared by the President, Ready Reserve units and individuals not assigned to units may be ordered to active duty for not more than 24 consecutive months.[6] Not more than 1,000,000 members of the Ready Reserve may be on active duty at any one time under this provision.

---

equipment from multiple units while drawing manpower primarily from a single unit. Research needs to determine where the problems of integration detract from a unit's ability to perform its mission. However, leadership, personality, and shared experience may play the key role in successful integration at the tactical unit level.

[6]The Ready Reserve consists of all reservists except those in an inactive or retired status.

- When the President determines it is necessary to augment the active force, Selected Reserve units and individuals not assigned to units and a special mobilization category of the Individual Ready Reserve may be ordered to active duty for not more than 270 days.[7] Not more than 200,000 members of the Selected Reserve may be on active duty under this provision at any one time, of whom not more than 30,000 may be members of the Individual Ready Reserve.

The provision to recall the Selected Reserve for periods up to 270 days was a liberalization, in 1994, of previous law that allowed a call-up of 90 days plus a 90-day extension. The provision for a special mobilization category within the Individual Ready Reserve was added in 1998.

These provisions limit the availability of reservists in several ways. Limits on the duration of call-ups and on the numbers allowed on active duty are such that the AC must be used to meet many contingencies, especially those in which declaration of a national emergency is deemed inappropriate or undesirable. Although it is only natural that the AC should be the first option considered to meet most contingencies, decisionmakers should consider the legal limitations on RC availability in determining the force mix. In general, the total force should be sized so that the AC can meet all but 200,000 manpower requirements in a peacetime contingency or 1,000,000 manpower requirements in a national emergency short of declared war.

Another limitation on availability is the requirement that reservists who are organized and trained in units must be recalled with their units rather than individually. As a practical matter, relatively small unit packages (unit type codes, or UTCs) can be specified in mobilization plans. Still, this provision could impede the flexible and efficient use of reservists in some circumstances.

---

[7]The Selected Reserve consists of individuals who participate in inactive-duty training periods and annual training. The individual Ready Reserve consists of Ready Reserve members who are not in the Selected Reserve. The special mobilization category of Individual Ready Reservists subject to call-up under the provision cited here must be within 24 months of separation from active duty, volunteers for entry into the special mobilization category, and in grades and skills designated by the service secretary concerned.

Availability When Not Mobilized

In his FY 1999 Air Force Posture Statement to Congress, Acting Secretary of the Air Force F. Whitten Peters stated that the service had helped to reduce operational tempo rates for active units through the

> creative use of the Reserve and Guard units and increases in manning in highly stressed specialties. However, these efforts have placed a new burden on the RC. During 1997, RC aircrews served an average of 110 days in uniform, with their support teams serving 80 days. (Department of the Air Force, 1998, pp. 2–3.)

Although those 110 days in uniform were not all days deployed overseas, they were days that may have been spent away from a full-time job, home, and family. Assuming that the RC is currently responding at or near its capacity to alleviate heavy deployment demands, this number of days in uniform may be an approximate upper limit on participation for the part-time RC force. Given time required for training and other administrative needs, availability of part-timers for deployment is considerably less than the 110 days in uniform stated above.

Thaler and Norton (1997) estimated the number of contingency deployment days available for AC and RC aircrews. Assuming a desired maximum of 120 temporary duty (TDY) days per year is established for active aircrews, they calculated that AC aircrews in the continental United States spend around 50 TDY days per year participating in individual training, joint exercises, and other activities not related to contingency operations. This leaves 70 days available for contingency operations.[8] Thaler and Norton also postulate that RC crews are available for 50 days of TDY per year, of which 15 days are available for overseas contingencies after noncontingency training, exercise, and other needs are satisfied. This limit has been validated through our own interviews with RC representatives at the headquar-

[8]Thaler and Norton postulate that aircrews in Europe require slightly more noncontingency TDY (60 days), leaving slightly fewer days available for contingencies (60 days).

ters and unit level.[9] AC aircrew deployments to Southwest Asia have been reduced to 45 days, which may be close to a minimum rotation duration to permit effective continuity of operations for fighters or command and control aircraft. Although a 45-day overseas deployment fits within the postulated 50-day limit for RC aircrews, such a deployment would not allow enough additional TDY for a RED FLAG or other unit training deployment after return to the home station. RC units can rotate aircrews to deployment sites in increments of fewer than 45 days, but the impact of shorter rotations on operational effectiveness remains to be examined.

The ANG has collected aircrew activity data that help to put Thaler and Norton's postulated level of availability in perspective. Figure 4.2 shows the number of TDY days experienced by active flying aircrews in ANG wings during FY 1997. The average number of TDY days is 37 days, somewhat fewer than the 50 days postulated by Thaler and Norton (1997).

TDY by reservists is included within total time spent in uniform. Figure 4.3 shows the number of days in uniform for active flying aircrews in ANG wings (all pay periods for traditional guardsmen and periods beyond the normal duty day for full-time technicians and active Guard/Reserve status individuals). We supply these data to help put the TDY data in perspective.

There is evidence that length of a TDY assignment is as important as the total yearly time away in determining the availability of RC aircrews. AFR volunteer rates for normal operations, small contingencies, and large contingencies during 1996 are shown in Figure 4.4. The data show a steep dropoff in volunteers for TDYs of 10 days or more. According to these data, the reserve strategic airlift pilot, who is offered shorter TDYs, may be more likely to volunteer than a fighter pilot or civil engineer who must volunteer in chunks of 45 days or more.

---

[9]Interviewees have said that for some personnel, 50 days per year is at the high end of what they can expect to provide given their full-time work and family responsibilities. However, except as constrained by law, full-time reservists may have the same availability as AC members.

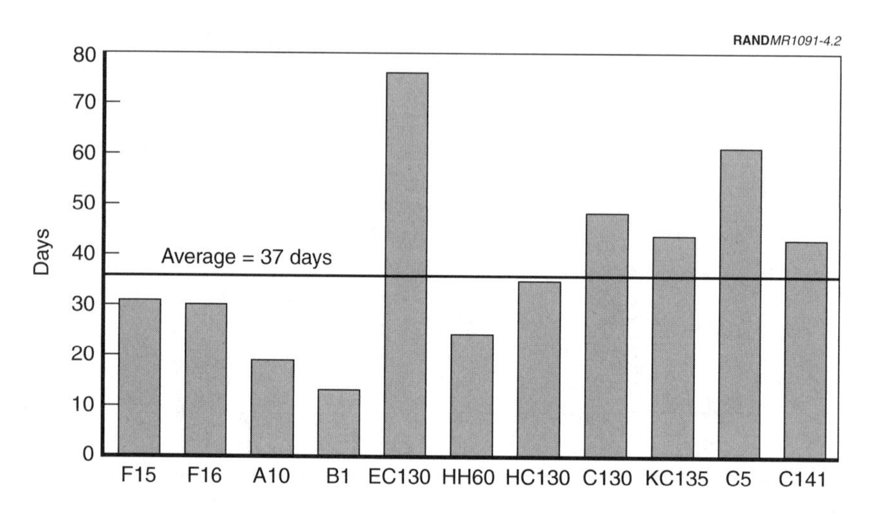

SOURCE: ANG Support Center Web site (http://xod.ang.af.mil/XODHome/Navigation/OpTempoIndex.htm).

Figure 4.2—FY 1997 TDY Days by ANG Aircrews

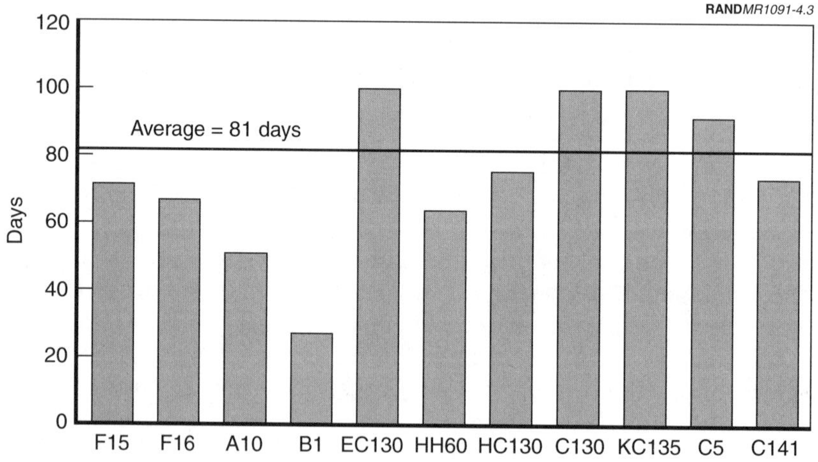

SOURCE: ANG Support Center Web site (http://xod.ang.af.mil/XODHome/Navigation/OpTempoIndex.htm).

Figure 4.3—FY 1997 Days in Uniform by ANG Aircrews

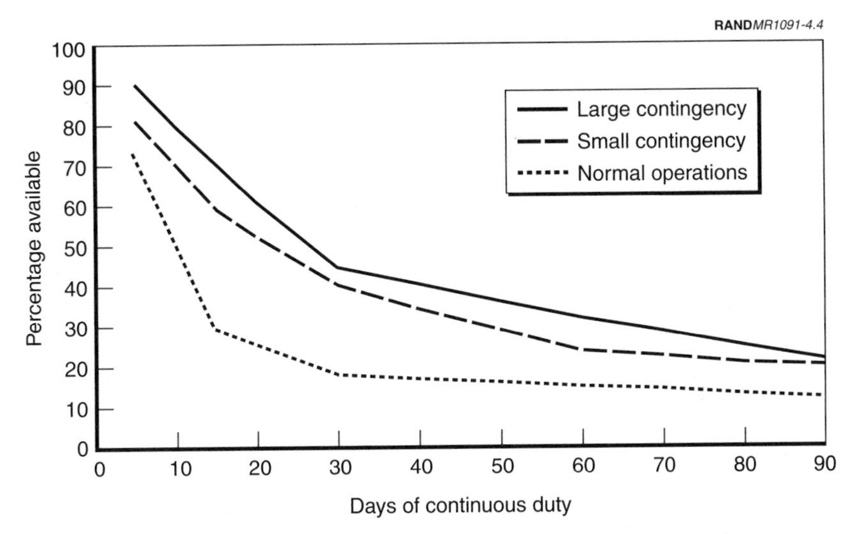

SOURCE: Air Force Reserve Command, *Air Force Reserve Review*, 1996.

Figure 4.4—Availability of AFR Aircrews by Duration of TDY

IMPLICATIONS FOR THE FORCE MIX

The evidence indicates that RC units are as ready as AC units for use in a major theater war, and have approximately the same availability, assuming mobilization. RC units maintain a high level of readiness, notwithstanding fewer training missions than AC units, because they have higher experience levels. However, for major operations short of declared war and for current peacetime contingency operations, RC units have limited availability relative to AC units.

Given readiness parity, 100 percent of the force could be in the RC with no loss of effectiveness, as indicated by the locus of the *readiness* constraint in Figure 4.5. This finding offers no guarantee that the RC could maintain its current high state of readiness in the long run if it had no AC as a source of experienced accessions. This *personnel flow* constraint will be considered separately in Chapter Five.

Availability, on the other hand, is limited for the RC because of its predominantly part-time workforce. Because of limits on mobiliza-

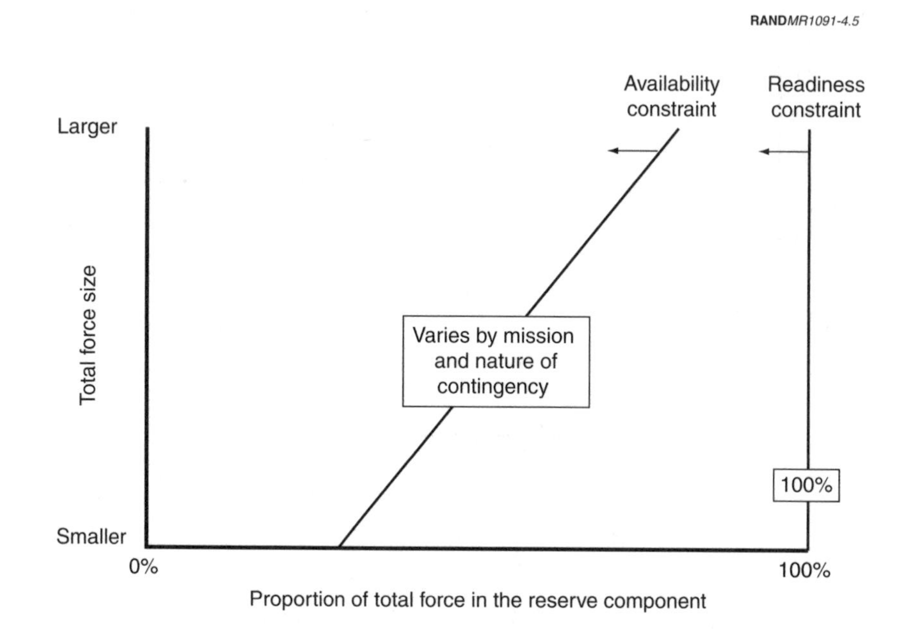

Figure 4.5—Locus of Availability and Readiness Constraints on the Force Mix

tion, the RC cannot satisfy certain short-notice or long-duration requirements specified in operational plans. Greater limits on cumulative deployment time and duration of deployment relative to the AC mean that the RC is less available for meeting contingency requirements. As the total force declines in size, assuming these force employment demands remain constant, the *availability* constraint, as depicted in Figure 4.5, permits a decreasing proportion of the total force to be placed in the RC.[10] However, the specific locus of this

[10]To illustrate why the line is sloped for major contingencies, consider a hypothetical MTW requirement for 2,000,000 military personnel, of whom a maximum of 1,000,000 may be mobilized reservists. If the total force consists of 3,000,000 military members, at least 1,000,000 must be in the active component in order to meet the MTW requirement. This limits the RC to no more than 67 percent of the total force. If the total force were larger—4,000,000—the requirement could be met with an RC of up to 75 percent of the total. To illustrate why the line is sloped for smaller peacetime contingencies, consider a hypothetical fighter force mix. Assume that AC squadrons can

constraint varies by mission or function. The availability constraint for a high-tempo asset such as an AWACS unit will be far different from the constraint for a low-tempo asset such as an air defense or space operations unit.

---

support 1200 deployed aircraft-days per year and RC squadrons can support 360. (These availability estimates are derived in Chapter Six.) If the total force must be sized at 20 fighter wing equivalents (FWEs), each consisting of three standard-sized squadrons, to support a two-MTW scenario and must supply, say, 50,000 deployed days per year, it can consist of, at most, 8.73 RC FWEs, or 43.6 percent of the total fighter force. The computations are:

RC contribution

8.73 FWE × 3 squadrons × 360 deployed aircraft-days = 9428 deployed aircraft-days

AC contribution

(20 – 8.73 FWE) × 3 squadrons × 1200 deployed aircraft-days = 40,572 deployed aircraft-days

The sum of the RC and AC contributions to deployed aircraft-days is 50,000. Any substitution of more RC FWEs for fewer AC FWEs results in fewer than 50,000 deployed aircraft-days.

If the total force is reduced to 16 FWEs but faces the same deployment demand, it can consist of, at most, three RC FWEs, or 19 percent of the total fighter force.

Chapter Five

PERSONNEL FLOW

Individuals who separate from active duty and subsequently affiliate with a reserve unit provide a significant base on which to build RC readiness. Years of experience gained by individuals in the AC and RC are not equivalent, because of differences in the time available to practice military skills: 38 days a year in the RC (62 days for pilots) and 225 days in the AC (RAND, 1992).[1] It is difficult for reservists who have never been on active duty to gain proficiency in complex skills with this limited amount of annual training.[2] Thus, RC readiness depends on a significant flow of experienced manpower from the AC to the RC, which imposes a constraint on force-mix planning.

EARLIER VIEWS ON PERSONNEL FLOW

The Gates Commission (formally, the President's Commission on an All-Volunteer Armed Force), formed in 1969 to make recommendations about the transition from conscription to an all-volunteer force, recognized the possible implications for RC sustainability. Among other concerns, the Commission examined whether an all-volunteer active force would sustain a sufficient flow of prior service (PS) accessions to the reserve forces and whether the remaining nonprior service (NPS) accession requirement could be met from the civilian

[1]The 38-day figure for reservists is based on 12 monthly weekend drills of two days each plus 14 days of active-duty training per year. Aircrews are authorized an additional 24 days of flight training periods, for a total of 62 days.

[2]This would not be true, however, for reservists whose full-time occupations closely match their reserve duties.

recruiting pool (President's Commission, 1970, pp. 109–117). With expected active and reserve force sizes, the Commission projected fewer active force losses and therefore a smaller pool from which to recruit PS accessions to the RC. However, with expected gains in reserve force retention (predicated on the Commission's recommended pay enhancements), the Commission concluded that the all-volunteer force would yield a ratio of PS to NPS reserve accessions at or above historical (pre-Vietnam) levels.

NPS accessions fell dramatically between 1970 and 1976, forcing the reserves to rely even more heavily than anticipated on PS recruiting (Brinkerhoff and Grissmer, 1986, p. 214). Shortfalls in NPS accessions caused reserve strengths to drop significantly from 1973 to 1978, during the early years of the all-volunteer force. By 1983, however, strength levels were restored and reserve forces continued to enjoy a high proportion of PS to NPS accessions. The PS proportion of total reserve accessions (across all services) was 31.9 percent in 1970, rising to a peak of 79.6 percent in 1974, and leveling out at 56.6 percent in 1982 (Brinkerhoff and Grissmer, p. 209).

Recognizing the importance of PS personnel to the reserves, Congress enacted the Army National Guard Combat Readiness Reform Act of 1992, which directed the Secretary of the Army to establish "an objective of increasing the percentage of qualified prior active-duty personnel in the Army National Guard to 65 percent, in the case of officers, and to 50 percent, in the case of enlisted members, by September 30, 1997."[3] An examination of relevant 1989 data showed that the Army National Guard was experiencing PS accessions well below these goals. However, had these goals applied to the ANG and AFR, they would have been far exceeded for both officers (85 and 87 percent for the ANG and AFR, respectively) and enlisted personnel (63 and 75 percent, respectively) (RAND, 1992, p. 263).

The active-to-reserve flow of pilots is especially meaningful to air RC readiness. Because of the length of undergraduate pilot training and subsequent weapon system qualification, most pilots in the air RC have prior active service. An analysis of DoD's Base Force (a plan-

[3]Section 1111 of Public Law 102-484, as amended by Public Laws 103-35, 103-60, and 103-337. See 10 USC 10105.

ning and programming template developed in 1989) revealed that it contained an active-to-reserve pilot ratio of about 1 to 0.9 (RAND, 1992, p. 234). This ratio was considered sufficient to support the required flow of pilots into the RC. However, previously published research does not indicate how far this ratio can shift toward a greater reserve proportion without jeopardizing the supply of PS pilots to the RC.

MODELING THE PERSONNEL FLOW CONSTRAINT

The active-to-reserve flow of personnel can be modeled by representing the demand for PS accessions in the RC and the availability of experienced losses from the active force. To compute the reserve demand, let A_r be the annual accession requirement for the reserve component, S_r the reserve component strength, and l_r the annual aggregate loss rate from the reserve component.[4] Assuming constant strength from year to year, the annual reserve accession requirement is a product of the loss rate and the strength:

$$A_r = (l_r)(S_r). \tag{1}$$

This accession requirement can be computed for the entire reserve strength or for any subset of it. For example, the requirement might be computed separately for officer and enlisted personnel or for a single occupational group such as pilots. The loss rate used for this purpose should, of course, correspond to the specific strength subset of interest.

The total pool of losses from active strength can be similarly computed by applying a loss rate to the active strength. Let P be the PS pool of AC separatees available for accession to the RC, S_a the active strength (either total strength or some subset of interest), and l_a the aggregate loss rate from the active force:

$$P = (l_a)(S_a). \tag{2}$$

[4]Population variables are expressed in upper case and rate or ratio variables are expressed in lower case.

Some active-force losses may be either ineligible for entry into the RC (e.g., retirees) or will not meet RC requirements (because of grade, years of service, or other similar characteristics). Of those eligible and meeting requirements, some will not be interested in affiliating with an RC unit. The PS pool must be adjusted to account for these factors. Let e represent the proportion of the PS pool eligible and meeting requirements and f the proportion willing to affiliate with a reserve unit. The adjusted available PS pool, AP, is thus:

$$AP = (e)(f)(P) = (e)(f)(l_a)(S_a). \quad (3)$$

The ratio of available PS recruits to accession requirements, AP/A_r, is thus a measure of the ability of the active force to sustain the experience needs of the reserve force, and it is a function of active and reserve strengths, active and reserve loss rates, and the reserve eligibility and affiliation rates of active force losses.[5] In some cases, there are specified values for this ratio. For example, to meet the congressional guideline discussed above, the ratio for Army National Guard accessions has a value of 0.65. The AFR plans for and obtains PS accessions for virtually all its pilot requirements, so that the ratio has a value of approximately 1. The ANG absorbs a larger proportion of NPS pilot accessions, resulting in a ratio of about 0.57.[6] Let x represent the desired AP/A_r ratio. The relationship can be expressed formally as

$$\frac{AP}{A_r} \geq x, \quad (4)$$

where x is a value between 0 and 1.

---

[5]This formulation assumes, for simplification, that PS accessions affiliate with reserve units in the same year they leave the active force. In the steady state, the formulation holds without this assumption.

[6]According to data supplied by ANG/XO, the ANG loses about 350 pilots per year. The ANG gets 180 UPT slots per year to be filled by ANG members, yielding 150 UPT graduates per year after training attrition. The remaining 200 losses are replaced using PS accessions.

Force Mix Based on Strength

If a desired value for x in Eq. (4) is known, the equations for AP and A_r can be used to solve for the needed ratio of active to reserve strengths. Substituting the right-hand sides of Eqs. (1) and (3) for A_r and AP yields:

$$\frac{(e)(f)(l_a)(S_a)}{(l_r)(S_r)} \geq x. \tag{5}$$

By transformation, the minimum ratio of active to reserve strength, S_a / S_r, can be found:

$$\frac{S_a}{S_r} \geq \frac{(x)(l_r)}{(e)(f)(l_a)}. \tag{6}$$

Given a specified value of x and expected values of l_r, e, f, and l_a, the minimum S_a/S_r can be computed.[7] For example, using notional values $x = 1$, $l_r = 0.10$, $e = 0.7$, $f = 0.5$, and $l_a = 0.06$, S_a/S_r must be at least 4.8 (i.e., active pilot strength must be at least 4.8 times as great as reserve pilot strength). For a case more representative of current ANG experience, a value of $x = 0.57$ might be used, yielding a required AC to RC ratio of 2.7 to 1.

Force Mix Based on Fighter Wing Equivalents

The force mix is often discussed in terms of a ratio of active-to-reserve *personnel strengths.* If so, the calculations above are sufficient to express the sustainability constraint. In other contexts, the mix is expressed as a ratio of active-to-reserve *units.* Notably, the

---

[7]In general, expected values should be selected so they produce a conservative estimate of the required ratio. In this case, a higher S_a/S_r ratio is more conservative. Thus, a prudent analyst would select a value for l_r near the high end of historically observed reserve loss rates and values for f and l_a near the low end of their historically observed ranges.

force-structure mix is often given as a ratio of active-to-reserve fighter wing equivalents (FWE).[8]

To determine required FWE ratios based on experienced pilot needs, further computations are required. These additional computations will convert pilot strengths to the number of aircraft they can support. Since FWE are a linear transformation of the number of operational aircraft, a proportion that holds for aircraft will also hold for FWE. The additional input factors required are crew ratios and allowances for pilots in nonoperational positions. Let FS_{ij} be the force structure (number of aircraft) in active or reserve component i and weapon system j, S_{ij} be the pilot strength, c_{ij} the crew ratio, and o_{ij} the proportion of the pilot force in operational positions.[9] Then

$$FS_{ij} = \left(o_{ij}\right)\left(\frac{1}{c_{ij}}\right)\left(S_{ij}\right) \tag{7}$$

and

$$\frac{FS_{aj}}{FS_{rj}} \geq \left(\frac{o_{aj}/c_{aj}}{o_{rj}/c_{rj}}\right)\left(\frac{S_{aj}}{S_{rj}}\right). \tag{8}$$

Equation (8) is expressed as an inequality because, like Eq. (6), it denotes the minimum ratio needed to satisfy personnel flow considerations. The results can be summed across all fighter weapon systems to determine a total force structure mix:

$$\frac{FS_a}{FS_r} = \frac{\sum_j FS_{aj}}{\sum_j FS_{rj}}. \tag{9}$$

---

[8]A FWE consists of 72 operational aircraft.

[9]Operational positions include rated position identifier (RPI) 1 and 2 positions in operational squadrons. Nonoperational positions include all other pilot requirements, including RPI 1 and 2 positions in training units.

Using notional values $o_a = 0.6$, $c_a = 1.25$, $o_r = 0.8$, $c_r = 1.25$, and $S_a/S_r = 4.8$ (and assuming, for the sake of simplification, that these values are constant across all weapon systems), we can compute $FS_a/FS_r = 3.6$. Translating this ratio to a proportion, these notional factors would suggest a force-structure mix that is no more than 22 percent in the RC. With $S_a/S_r = 2.7$ (more appropriate for the ANG), the result is $FS_a/FS_r = 2$, or a force that is no more than 33 percent in the RC.

The required force-structure ratio can be further refined to recognize that reserve fighter pilot accession requirements can be met from sources other than active-duty fighter pilot losses. Some requirements can be met using losses from other services or pilots whose active-duty flying was not in fighters. Because these pilots, especially those who have not flown fighters, will require extensive transition training and will dilute experience levels in reserve units, a limit on the proportion of reserve fighter pilot accessions taken from these sources would appear reasonable. To see how this limit affects the force mix, we introduce another factor in Eq. (4). Let t be the maximum proportion of reserve fighter pilot accessions that may be transitioning from a different flying community. If *t* were set, for example, at 30 percent, then only 70 percent (1 – t) of the reserve accessions required to meet the PS fighter pilot accession target would have to come from the active-duty fighter pilot losses. Thus, if *AP* represents the available PS pool drawn from active-duty fighter pilot losses, Eq. (4) becomes

$$\frac{AP}{A_r} \geq (1-t)(x) \text{ and} \tag{10}$$

Eq. (6) becomes

$$\frac{S_a}{S_r} \geq \frac{(1-t)(x)(l_r)}{(e)(f)(l_a)}. \tag{11}$$

If the force-mix constraints are recalculated with t = 0.3, the strength ratio constraint, S_a/S_r, drops from 4.8 to 3.3 and the FWE ratio constraint, FS_a/FS_r, drops from 3.6 to 2.5 (28 percent in the RC). Using the PS accession ratio more representative of the ANG, S_a/S_r drops from 2.7 to 1.9 and the FWE ratio constraint, FS_a / FS_r, drops from 2 to 1.4 (42 percent in the RC).

Figure 5.1 captures these sustainability calculations graphically. It shows annual losses as a subset of total AC fighter pilot strength. Within the annual loss number is a subset of those eligible to affiliate with the RC. Within that population is a smaller number willing to affiliate. This number, plus similar eligible and willing pilot losses from other than the Air Force fighter pilot inventory, must be large enough to fill annual RC fighter pilot accession requirements.

Using the Model

The input values used for the above calculations are rough estimates developed from limited data sources or partially informed opinion. Determining more precise expected values for the various input factors needed for this computation is beyond the scope of this study. However, the factors can be determined through analysis of histori-

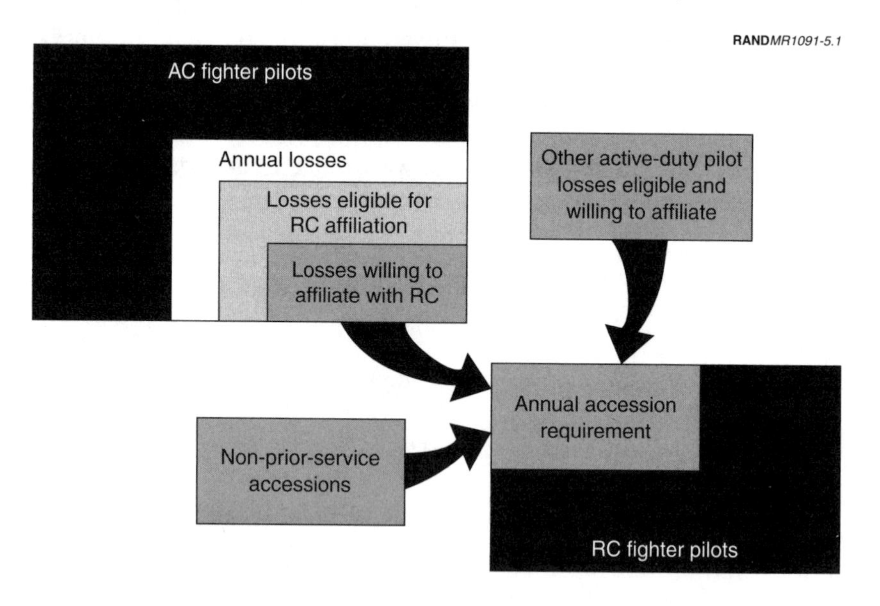

Figure 5.1—Sustainability of RC Fighter Pilot Requirements

cal personnel flows and expert judgment on experience requirements. We note that the Quadrennial Defense Review (QDR) recommended a reduction of the active component to about 12 FWE and an increase in the reserve component to 8 FWE. This yields an FS_a/FS_r ratio of 1.5 for fighter units, suggesting that a close examination of the sustainability constraint is warranted.

An aircraft-denominated force mix can be analyzed using requirements for military occupations other than pilot, given some figure comparable to the crew ratio that relates manpower to aircraft. However, for most other occupations, the RC can absorb an appreciable proportion of NPS accessions. For those occupations, the required ratio of prior-service recruits to reserve accession requirements (x in the equations above) will be much less than 1, as it is for AFR pilots, or perhaps even less than the 0.57 we assumed for ANG pilots. Thus, the pilot-based analysis is likely to provide the most tightly constraining result.

IMPLICATIONS FOR THE FORCE MIX

An upper bound on the proportion of the total force in the RC, related to personnel flow, can be determined if other related parameters are known. These other parameters are likely to vary by mission or MDS and also by differences between the ANG and AFR in their perceived ability to absorb inexperienced UPT graduates. Accordingly, as depicted in Figure 5.2, the constraint will also vary by mission or MDS and by component. In the notional examples provided here, the fighter force-structure constraint varied from 28 percent for the AFR to 42 percent for the ANG.

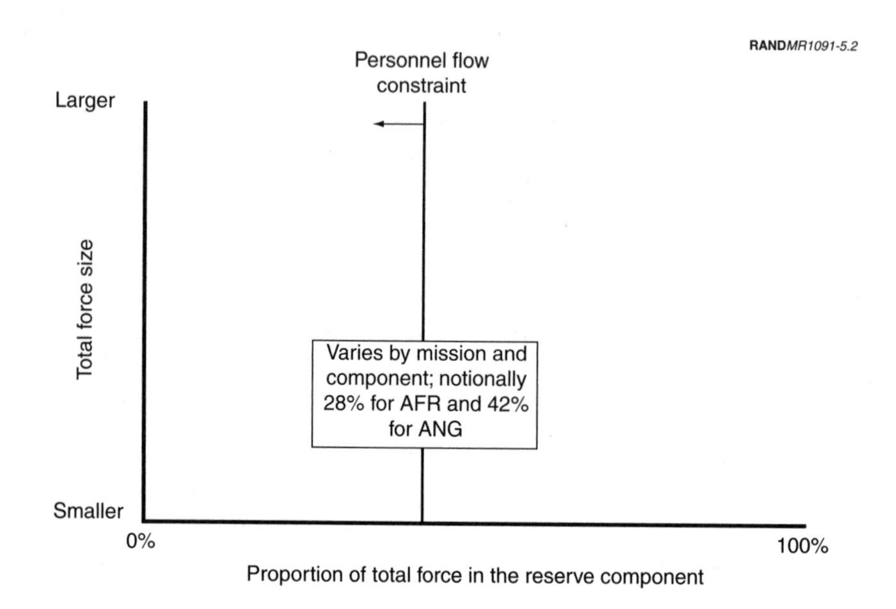

Figure 5.2—Locus of the Personnel Flow Constraint on the Force Mix

Chapter Six

COST

Past studies have shown that operations and support costs for RC flying units are generally, but not universally, lower than the costs of similarly equipped active units. A 1990 DoD report to Congress on total force policy compared the costs of 11 unit types found in both the AC and RC and found lower costs in the RC for all but one type of unit (DoD, 1990). Palmer et al. (1992, p. 49) found that costs in the RC are lower than those in the AC largely because RC units use less full-time manpower and fly fewer hours per year.

In this chapter, we will demonstrate that using total operations and support costs for AC and RC units is not sufficient in itself to indicate the relative cost advantages of one component over the other. Unit outputs must also be considered. The appropriate basis for comparing costs among components is cost per relevant output.

Relevant outputs will vary in different contexts. For meeting the demands of major theater wars (MTWs), where full mobilization of reserve forces can be assumed, the relevant output is a trained and ready unit. In this context, total operations and support costs per unit are appropriate for making AC/RC comparisons. This is the context and the approach used in most AC/RC cost comparisons. For meeting the demands of small-scale contingencies (SSCs)—a high-tempo context with more or less continuous operations that fall below the threshold for mobilization of reserve forces—relevant output will be mission-dependent. For fighter units, for example, it might be the number of deployed aircraft-days per unit of time that a unit can support. For airlift units, it might be flying hours devoted to productive (i.e., moving freight and passengers), as opposed to training, missions. We will suggest that the least costly force mix for meeting MTW demands may not be the least costly mix for meeting ongoing SSC demands. Thus, for a given budget constraint, force-mix decisionmakers may have to make tradeoffs.

For several reasons, this chapter contains more details than the preceding chapters. First, in addressing cost, we often saw significant differences in perspective between the AC and the RC. To counter subjective estimates of relative AC and RC costs, we rely on concrete information. Second, as mentioned above, we propose new approaches for evaluating costs in light of the kinds of demands generated by SSCs. We thought that concrete examples would help to make the case for these new approaches.

MEETING MTW DEMANDS: FORCE-STRUCTURE AVAILABILITY

Conventional approaches to comparing AC and RC costs for air forces have generally attempted to capture the operating costs of similarly sized and equipped squadrons. The implicit premise of these cost comparisons is that a reserve squadron operating a given MDS is equivalent to an active squadron operating the same MDS for some cost-related purpose. That purpose, presumably, is utilization of the unit in an MTW, when the option to mobilize reserve units makes them fully available during a period of some duration. Thus, to meet MTW demands, the cost of providing a trained and ready unit is the peacetime operating cost of the unit.

It may be useful to discuss these costs on a per primary aircraft authorized (PAA) basis rather than a per-unit basis, because units of the same MDS type often vary in the number of PAA they are assigned. Another refinement, pursuant to our recommended cost-per-output approach, would consider designed operational capabilities (DOCs), which can also vary across units of the same type. Some units are, by design, more capable than others.

Cost Comparison Complexities

Cost comparisons between active and reserve units are quite complex. Cost analysts must classify costs as direct versus indirect, fixed versus variable, and recurring versus transitional. These distinctions may be understood as follows:

- Direct costs are incurred within the activity being costed, whereas indirect costs are either overhead costs or costs incurred by other activities to support the costed activity. For a flying squadron, direct costs include pay and other personnel costs of individuals assigned to the squadron; petroleum, oil, and lubricants (POL) consumed by the squadron; and other operations and maintenance costs, such as consumable supplies and parts. Indirect costs include headquarters staffs of larger commands to which the unit is assigned, accession and training costs needed to sustain the personnel strengths of the unit, depot maintenance, medical support, and other base support.
- Fixed costs are those that would be incurred whether or not the unit is in the force structure, whereas variable costs are those that are incurred only if the unit is in the force structure. All direct costs are variable, whereas indirect costs have fixed and variable components. For example, headquarters staff sizes are unlikely to be affected by the addition or subtraction of a given unit.
- Marginal costs are those fixed and variable costs that are incurred as a result of a unit being part of the force structure or, conversely, those costs that could be removed from the USAF budget without affecting any other organization if the unit under analysis were disestablished.
- Total costs include the marginal costs plus a proportional cost of the support structure allocated from the overall USAF overhead costs, such as headquarters, the acquisition organizations, and medical organizations. Total cost development is the objective of *activity-based costing*, which is receiving much attention but for which a straightforward methodology is not available to analysts making force-structure adjustments. This approach is used for setting DoD reimbursement rates for various outputs, such as airlift or sealift.
- There is a recurring, or steady-state, cost level for ongoing operation of a unit. However, establishing or disestablishing a unit will entail some transitional costs or savings.

There is no one combination of these cost categories that is appropriate for all purposes. For example, when comparing the cost of alternative force structures, such as placement of a unit in the AC or RC, analysts should compare the marginal cost of an AC or RC unit, including direct and indirect variable costs and excluding fixed costs. In practice, it is often difficult to identify and exclude the fixed component of indirect costs, resulting in overstating marginal costs. However, if the cost analyst is trying to determine the full cost of a unit, such as for setting a reimbursement rate to recover costs from another agency, fixed costs should be allocated to the unit and included in the cost estimate. For quick comparison of many options, analysts must generally confine their attention to recurring costs only. However, as the options become concrete, the proper approach is to consider transitional costs as well. This requires analysts to develop costs for a series of post-decision time periods and to discount them to the decision point. However, transitional costs, especially for indirect costs, may be difficult to determine.

To add to the complexity, categorizing costs as direct/indirect, fixed/variable, and marginal/total depends on the *unit of analysis.* If one is considering the marginal cost of adding or deleting squadrons from the AC or the RC, the *squadron* is the unit of analysis. All costs incurred within the squadron are regarded as direct, variable costs. At a less-aggregate level, one could consider increasing or decreasing the number of PAA in a squadron. In that case, the *PAA* would be the unit of analysis. Certain squadron overhead costs would be regarded as fixed; costs related to the number of PAA in the unit would be regarded as direct and variable. At an even less-aggregate level, the number of *flying hours* per PAA could be the unit of analysis. Certain ownership costs of a PAA would be regarded as fixed; costs related directly to flying hours would be regarded as variable. In this study, we generally treat the squadron as the unit of analysis.

Cost Comparison Results

Several relatively recent studies by DoD (1990), the Institute for Defense Analysis (IDA) (Wilson et al., 1992), and RAND (Palmer et al., 1992) have compared the costs of selected unit types in the AC and RC. Findings are shown in Table 6.1. It is not clear that the authors of these studies have succeeded in isolating the marginal costs of

Table 6.1

Recurring Peacetime Costs for Selected Unit Types—Previous Cost Comparison Studies
(Costs in $millions/annual flying hours [FH])

Unit		Crew		Active		ANG		AFR	
Type	PAA	Ratio	Source	Cost	FH	Cost	FH	Cost	FH
F16	24	n.s.[a]	DoD[b]	64.8	8134	48.8	5064	51.6	4682
F16C/D	24	n.s.	IDA[c]	91.7	8134	58.7	5064	62.4	4682
F16	24	n.s.	RAND[d]	63.6	unk	36.0	unk	38.4	unk
KC135	10	n.s.	DoD	36.3	2840	44.2	3500	42.1	3801
KC135	10	1.27	IDA	47.5	2840				
KC135	10	1.5	IDA			51.1	3500	54.2	3801
KC135	10	n.s.	RAND	35.0	unk	30.0 (tenant)	unk	32.0 (tenant)	unk
KC135	10	n.s.	RAND			34.5 (host)	unk	48.0 (host)	unk

[a]n.s. = not specified; unk = unknown.
[b]Source: DoD (1990), Table 5. Costs indicated are in FY 1992 dollars.
[c]Source: Wilson et al. (1992). This paper is related to DoD (1990). It includes the same direct and average annual equipment costs plus infrastructure costs.
[d]Source: Palmer et al. (1992). Costs indicated are in FY 1993 dollars.

unit operation, or even that they intended to do so in every case. Almost certainly, these estimates include some allocations of fixed costs.

Although there appears to be some consistency among the costs reported by various studies, there are significant similarities and differences in the underlying data. The DoD and IDA studies were companion pieces.[1] The DoD total cost figures are composed of direct unit costs, defined as personnel costs, operating costs, and something akin to a depreciation cost for unit equipment. The IDA total cost includes these elements plus an allocated infrastructure cost. The RAND figures include direct costs such as personnel and consumables, but also depot maintenance—an item that the IDA report

[1]The DoD study group did some of its own analysis and commissioned several supporting studies by federally funded research and development centers whose reports were to be published separately. The IDA study was one of those.

includes in infrastructure costs—and indirect costs such as base operating support and training.

These are peacetime costs. In the event of an MTW, mobilization of RC units would make their pay and flying hours comparable to AC units and higher operating tempos would drive up the costs of both AC and RC units.

Past studies have generally shown that RC units, flying fewer hours and relying to a significant extent on part-time labor, are less costly than AC units. The F16 data shown in Table 6.1 are typical of these studies. The KC135, also shown in Table 6.1, is an anomaly—in these earlier studies, RC KC135 units were generally found to fly more hours and therefore to be more costly than AC units.

The office of the Deputy Assistant Secretary of the Air Force for Cost and Economics (SAF/FMC) maintains a unit costing model and data base called SABLE (for Systematic Approach to Better Long Range Estimating) that can be exercised to obtain current cost estimates. Data obtained from the SABLE model are shown in Table 6.2.

All costs shown in SABLE are variable—varying as a function of PAA, flying hours, or authorized personnel strengths. However, it is likely that some of the underlying factors (such as installation support costs per person) have allocated fixed costs embedded within them.[2]

The SABLE data show that RC crews generally fly fewer hours than AC crews, and in some cases have lower crew ratios, resulting in lower annual O&S costs per PAA for the RC. (As with earlier studies, these data show that the KC135R is an exception to the general rule.)

---

[2]During preparation of this report, some RC representatives we interviewed were skeptical that past studies or the SABLE model accurately capture the cost advantages (a part-time workforce and a less-elaborate support structure) inherent in RC operations. We note, however, that SABLE cost factors are designed to take such differences into consideration. Personnel counts and costs are differentiated for active-duty and drill, rated and nonrated, officer and enlisted, and military and civilian personnel. The per-capita installation support cost factor is higher for AC units. Active personnel incur several categories of cost (permanent change of station [PCS], medical) not present for RC personnel. We find no ostensible basis for discrediting SABLE-generated cost comparisons.

Table 6.2

Recurring Peacetime Costs for Selected Unit Types—SABLE Model (FY 1997 $)

Unit Type	Component	PAA	Crew Ratio	Annual Flying Hours (FH)	FH per Crew	Annual O&S Costs	Annual O&S Cost per PAA	Cost per FH
F16C	Active	18	1.25	6,426	286	$46.1M	$2.6M	$7,174
F16C	ANG	18	1.25	4,230	188	$32.7M	$1.8M	$7,730
F16A	AFR	15	1.25	3,689	197	$31.1M	$1.2M	$8,432
F15C	Active	18	1.25	5,670	252	$64.6M	$3.6M	$11,393
F15E	Active	18	1.25	6,246	278	$69.9M	$3.9M	$11,191
F15A	ANG	18	1.25	3,888	173	$46.7M	$2.6M	$12,011
A10	Active	12	1.5	5,304	295	$31.2M	$2.6M	$5,882
A10	ANG	18	1.25	4,770	212	$34.0M	$1.9M	$7,128
A10	AFR	12	1.25	2,993	199	$25.5M	$2.1M	$8,520
C130E	Active	16	2	10,304	322	$58.2M	$3.6M	$5,648
C130E	ANG	8	1.75	3,120	223	$25.6M	$3.2M	$8,205
C130E	AFR	8	1.75	2,841	203	$22.2M	$2.8M	$7,814
C141B	Active	16	1.8	16,192	562	$103.1M	$6.4M	$6,367
C141B	ANG	8	2	2,928	183	$34.8M	$4.3M	$11,885
C141B	AFR	8	2	2,772	173	$31.9M	$4.0M	$11,508
C141B	AFR (Assoc)	16	1.8	4,990[a]	173	$23.9M	$1.5M	$4,790
KC135R	Active	12	1.27	3,672	241	$29.3M	$2.4M	$7,979
KC135R	ANG	10	1.5	3,500	233	$35.9M	$3.6M	$10,257
KC135R	AFR	10	1.27	2,940	231	$27.3M	$2.7M	$9,286
KC10	Active	12	2	7,164	299	$64.7M	$5.4M	$9,031
KC10	AFR (Assoc)	12	1.5	3,740	208	$41.8M	$3.5M	$11,176

NOTE: Data were developed from the SABLE cost model maintained by SAF/FMC. PAA, crew ratios, flying hours, and operating and support (O&S) costs were extracted from SABLE. Three of the ratios reported in the table (flying hours per crew, cost per PAA, and cost per flying hour) were computed for this analysis using data extracted from SABLE. The cost per PAA and cost per FH reported here are total unit costs per PAA or FH. They should not be confused with SABLE logistics cost factors that are expressed on a per-PAA or per-FH basis.

[a]SABLE indicates zero flying hours for a C141B associate unit. To calculate the cost per flying hour, we assumed that AFR associate unit crews would fly the same number of hours as AFR non-associate unit crews.

However, costs per flying hour are generally lower for AC units, perhaps because of economies of scale.

Differences in Designed Operational Capabilities

RC units in some cases have narrower DOCs than comparably equipped AC units. The differences in DOCs are necessary because the RC units, with fewer annual flying hours per crew, are unable to train adequately to the wider range of capabilities found in the AC units. Ideally, a cost-per-output approach to cost comparisons would account for these differences in capability in the context of providing ready units to meet MTW demands. However, we were unable in the scope of this research to develop an approach for computing a *cost per capability.* Capabilities are not uniform in many important respects. For example, some are more costly to develop than others and some may be more valuable to end users (warfighting commands) than others. An explicit cost-per-capability measure would have to account for this lack of uniformity. Lacking such an approach, decisionmakers considering alternative AC/RC mixes must subjectively weigh relative costs and capabilities of AC and RC units, particularly in the face of the often differing capabilities among various MDS of like-mission aircraft.

A Strategy for Optimizing Force-Structure Availability and Cost

Since the annual operating costs of RC units are generally less than those of AC units, cost-effectiveness in meeting MTW demands is achieved by placing as much of the force structure in the RC as possible. For fighter units, force structure is generally measured in FWE, calculated as the number of PAA in the inventory divided by 72 (the number of aircraft in a notional wing consisting of three 24-PAA squadrons). The force structure is optimized by placing just enough force structure in the AC to meet requirements that are incompatible with RC employment and placing the remainder of the force mix in the RC. The force mix at the end of FY 1997 was 13 AC FWE and 7 RC FWE, but the Quadrennial Defense Review determined that the mix could shift to 12 AC FWE and 8 RC FWE (Cohen, 1997, p. 30).

MEETING SSC, OOTW, AND OTHER PEACETIME DEMANDS

In addition to providing ready forces for MTWs, AC and RC units provide forces for SSCs and OOTW. In the past, the force structure required to meet these non-MTW demands was considered to be a lesser included case within the MTW-related force structure. However, it is becoming apparent that operating tempos imposed by SSCs and OOTW are placing great stresses on the current force structure. The force structure that is best for meeting MTW demands may not be best for meeting SSC and OOTW demands. Accordingly, it is appropriate to compare the relative costs of AC and RC units to meet these non-MTW demands.

In anything other than an MTW scenario, widespread mobilization of RC units is unlikely. Thus, as discussed in Chapter Four, RC availability to generate productive output in support of SSC and OOTW demands is more limited than that of the AC. For cost-comparison purposes, the appropriate cost-per-output approach is to divide the annual operating cost by a denominator that represents a productive operational output in an SSC (nonmobilization) environment. For example, the denominator for fighters might be the maximum number of days deployed for contingency operations that the unit can be expected to support. For airlift units, the denominator might be productive flying hours. In the following paragraphs, we develop examples of how these cost comparisons might be developed for fighter and airlift assets. Similar peacetime output measures and cost comparisons could be constructed for other missions and weapon systems.

Measuring Peacetime Fighter Deployment Capacity and Its Cost

As discussed in Chapter Four, Thaler and Norton (1997) estimated that of 120 total TDY days available per year, AC fighter aircrews have 70 days per year available for contingency operations.[3] Thaler and Norton also postulate that RC crews are available for 50 days of TDY

[3]Thaler and Norton postulate that aircrews in Europe require slightly more noncontingency TDY (60 days), leaving slightly fewer days available for contingencies (60 days).

per year, of which 15 days may be used for overseas contingencies. For an entire squadron, the number of available days per crew is multiplied by the number of crews in the unit. For 24-PAA squadrons with a crew ratio of 1.25, this yields 1500 deployed air-crew-days per year for an AC unit and 450 for an RC unit.

Information about deployable aircrew-days can be combined with squadron operating costs and other factors to determine the cost of a deployed aircraft-day. Assuming that deployed units operate at the same crew ratio as at their home bases, it appears that AC units can provide 1200 days of deployed aircraft operations per year whereas reserve units can provide 360. (Available deployed aircrew-days must be divided by the deployed crew ratio to determine the number of available deployed aircraft-days.) Dividing the annual operating costs of a fighter unit by these figures yields a cost per deployed aircraft-day.[4] As recapitulated in Table 6.3, using the SABLE-reported F16C unit operating costs shown in Table 6.2, the cost per deployed aircraft-day for an AC unit is $51,250. The comparable cost for an ANG unit is $121,111.

Measuring Peacetime Airlift Productive Capacity and Its Cost

For airlift units, contingency-deployed aircraft-days may not be the appropriate measure of useful output. A more appropriate measure might be the cost per productive flying hour. In this measurement, productive flying hours are those devoted to Joint Chiefs of Staff (JCS)-directed missions, channel traffic, and special assignment airlift missions (SAAMs). Table 6.4 provides the cost calculations. Note that this table depends critically upon the assumption of the proportion of flying devoted to productive missions. We had no immediate access to empirical data as a basis of the rates used in Table 6.4. We relied, instead, on rough estimates supplied by a colleague familiar with active and reserve airlift operations. Unless these estimates are widely inaccurate, data suggest that AC units provide the least

[4]Annual operating costs of both AC and RC units might rise as a result of a prolonged contingency deployment for reasons such as increased flying hours, transportation of unit personnel and other assets to and from the theater of operations, and increased consumption of munitions. We base our cost comparisons on peacetime operating costs, but a more rigorous cost analysis might consider the likely changes in operating costs associated with typical OOTW and SSC deployments.

Table 6.3

Cost per Deployed Aircraft-Day, F16 Squadrons

		AC	ANG
1.	TDY days per year per crew member	120	50
2.	Training and other noncontingency TDY days required per year	70	35
3.	Available contingency TDY days per year (line 1 minus line 2)	50	15
4.	PAA	24	24
5.	Crew ratio	1.25	1.25
6.	Contingency-deployed days per year (line 3 × line 4 × line 5)	1500	450
7.	Contingency-deployed aircraft operation days per year (line 6 divided by line 5)	1200	360
8.	Unit operating costs per year (Table 6.2)	$61.5M[a]	$43.6M[b]
9.	Cost per contingency-deployed aircraft-day (line 8 divided by line 7)	$51,250	$121,111

[a]18-PAA unit cost ($46.1M) linearly scaled to 24-PAA.
[b]18-PAA unit cost ($32.7M) linearly scaled to 24-PAA.

Table 6.4

Cost Per Productive Flying Hour, C141 Squadrons

			RC	
		AC	Independently Equipped	Associate
1.	Flying hours per PAA per year (Table 6.2)	1,012	347	312
2.	Proportion productive (JCS-directed, channel, SAAM) (rough estimates)	80%	50%	50%
3.	Productive FH per PAA (line 1 × line 2)	810	174	156
4.	Annual cost per PAA (Table 6.2)	$6.4M	$4.0M	$1.5M
5.	Cost per productive FH (line 4 divided by line 3)	$7,901	$22,988	$9,615

expensive lift, followed by RC associate units. However, an associate unit must be paired with an AC unit. In contrast to these pairings, independently equipped airlift units cost far more per productive flying hour.

A Strategy for Optimizing Deployment Capacity and Cost

Providing the greatest capacity for SSC, OOTW, and other peacetime operations at a given budget level (or, equivalently, minimizing the cost for a given capacity) requires force mixes different from those that optimally meet MTW demands. To meet non-MTW demands, the optimal mix in fighter MDSs places just enough of the force in the RC to satisfy social and political considerations and the remainder in the AC. The optimal mix for meeting non-MTW airlift needs splits the force evenly into AC and RC associate units.

TRADEOFFS BETWEEN THE CONFLICTING DEMANDS

A force mix that optimally meets non-MTW demands (by minimizing the proportion of the force in the RC, subject to social and political constraints) may not provide a force structure large enough to meet MTW demands. Likewise, a force mix that optimally meets MTW demands (by maximizing the proportion of the force in the RC, subject to mobilization-related availability constraints) may not provide enough deployment capacity to meet non-MTW demands.

Conflicting Demands for Fighter Force Structure

When faced with a conflict between MTW and non-MTW demands, decisionmakers must weigh the tradeoffs. Figure 6.1 illustrates how the tradeoffs can be conceptualized and quantified for the fighter force. It provides information about potential alternative FWE force mixes with costs held constant at the level of the QDR-proposed mix of 12 AC and eight RC FWE. It shows that as the number of RC FWE increases (from zero to ten on the horizontal axis), the number of total force FWE available to meet MTW needs (on the left axis) also increases. In addition, as the number of RC FWE increases, the number of contingency-deployed aircraft-days that can be generated to meet non-MTW needs (on the right axis) declines. Note that the

QDR force mix, with eight FWE in the RC and a total of 20 FWE, can generate an estimated 51,800 contingency-deployed aircraft-days. Holding costs constant, each increase of one FWE in the RC results in a decrease of 0.7 FWE in the AC, or a net increase of 0.3 FWE in the total force. However, each increase of one FWE in the RC decreases the total force's capacity for deployed aircraft-days by about 1480 per year (1080 additional deployed days attributable to the additional RC FWE minus 2560 deployed days attributable to the 0.7 FWE lost from the AC). As an illustration of how the figure can be used, consider a force mix that contained only three FWE in the RC. Reading up to the FWE line and across to the left axis, it can be seen that the total force, if held to the same cost as the 20-FWE QDR mix, would provide only 18.5 FWE for use in an MTW, of which 15.5 would be in the active force. However, it would have a capacity for 59,200 deployed aircraft-days.

To develop the underlying data for Figure 6.1, we had to make an assumption about the marginal cost of an FWE in the AC and RC. To simplify our cost calculations, we assumed that marginal units would be equipped with F16Cs, costed as shown in Table 6.3.[5]

In practice, the linear changes in cost assumed here would probably hold over modest changes from the current force structure but not for more radical changes. At the extremes, some costs that are considered fixed for small changes would become variable. For example, if the current AC/RC mix were tilted much more toward the RC, sustainability of RC pilot accessions from AC trained pilot losses might become infeasible, forcing the RC to incur significant additional costs to train NPS pilots and fly them enough to reach proficiency in their weapon systems.

A Comparative Look at Airlift Force Structure

We have not analyzed MTW versus non-MTW demands for all MDSs. However, since a significant proportion of airlift capacity is in the RC, we have developed a notional approach for considering the tradeoffs

---

[5]An assumption of F16C equipage for the marginal unit is reasonable, because F16s are by far the most numerous fighter aircraft in both the AC and RC.

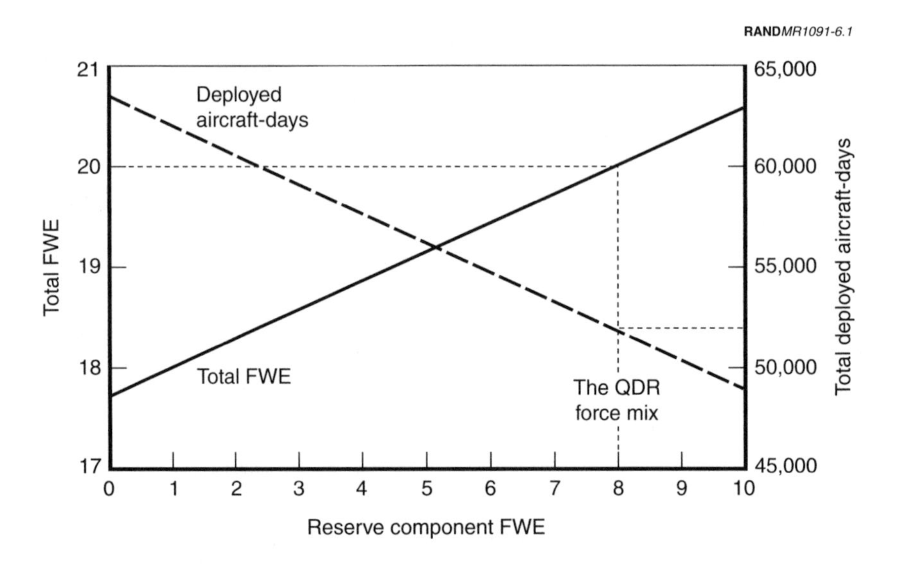

Figure 6.1—Alternative Equal-Cost Force Mixes

in an airlift MDS. We illustrate the approach using available data for the C141B.

The marginal cost of providing MTW capacity can be derived from Table 6.2. The data indicate that C141B RC units, like fighter units, are less costly than AC units, allowing more force structure to be generated for MTW purposes by shifting the AC/RC mix toward the RC. However, in airlift units, the cost differences between associate and independently equipped units require a more nuanced analysis. For this purpose, we considered two cases. In the first case, we pair an AC and RC associate unit and observe the total cost of the two units (associate units must be paired with AC units). In the second case, we determine the cost for providing the same number of aircraft in RC independently equipped units. In the first case, a 16-PAA AC C141B unit ($103.1M) paired with a 16-PAA AFR associate unit ($23.9M) has a total annual cost of $127M. The same force structure could be provided by two 8-PAA independently equipped RC units ($34.8M for an ANG unit or $31.9M for an AFR unit) at a total cost of $63.8M to $69.6M. Since the marginal cost of the force structure in the independently equipped units is lower than in the AC/associate unit pairing, greater equipment capacity is gained by shifting the force-structure mix toward independently equipped units. However, independently equipped units would provide only 32 crews, whereas the active/associate pair would provide 58 crews, permitting more intense operation of the available equipment.

For meeting non-MTW demands, Table 6.4 indicates that AC/ associate pairings provide more lift capacity than independently equipped RC units at a given budget constraint. Thus, there is a potential conflict between MTW and non-MTW demands if equipment capacity in an MTW scenario is more important than aircrew availability.

Making Tradeoffs

Making the tradeoffs between potential MTW capacity and realized peacetime capacity requires balancing the risks and benefits in MTW and non-MTW scenarios. The task is compounded by uncertainty about the demands for capacity in either scenario. Analysis can aid the decision process by providing estimates of expected demands and using them as a basis for quantifying expected risks and stresses (such work is beyond the scope of this analysis).

IMPLICATIONS FOR THE FORCE MIX

As depicted in Figure 6.2, cost considerations argue for a larger proportion of the total force in the RC when contemplating MTW scenarios and a smaller proportion when contemplating SSC and OOTW scenarios.

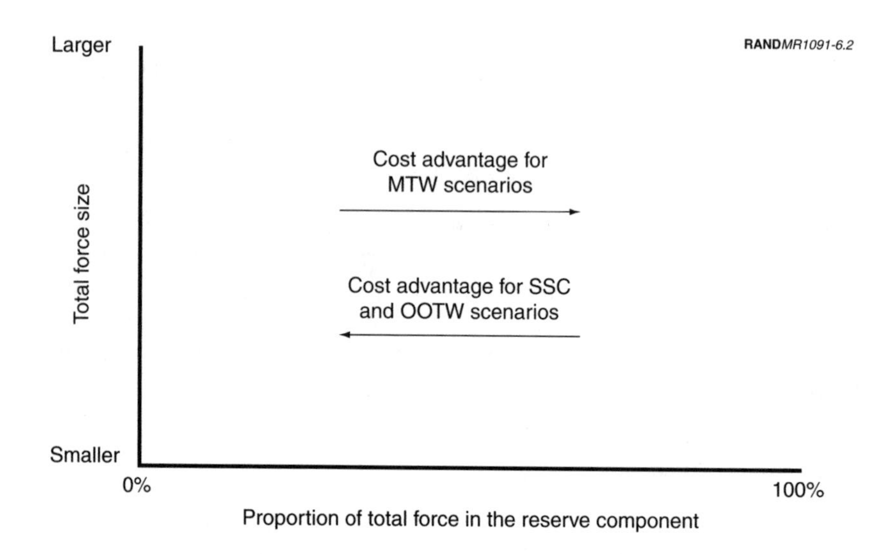

Figure 6.2—Cost Considerations in the Force Mix

Chapter Seven

CONCLUSIONS

Our model for graphically depicting a set of force-mix principles presented in Chapter Two postulated a feasible region within which a range of force mixes would be acceptable and within which cost considerations could prevail. In examining the available evidence, we found that the loci of some of these constraints are currently unknowable and that others are mission- or component-dependent. Where possible, we derived notional results using mission-dependent values pertinent to the fighter force.

The results shown in Figure 7.1 depict the ANG case, where notional personnel flow constraints might allow the RC to occupy up to 42 percent of the total fighter force. A feasible region is thus created to the right of the minority status constraint. The feasible region might be reduced if an availability constraint came into play or if a decisionmaker were to supply some judgmental locus for the social identification, embeddedness, and investment (IE&I) constraint. Within this feasible region, cost-conscious decisionmakers would gravitate toward a 42 percent mix if they were primarily concerned about preparedness for MTW scenarios or toward a 20 percent mix if they were more concerned with meeting current contingency deployment needs. It is possible, of course, to weigh cost more heavily than either the personnel flow or social constraints. In that case, decisionmakers might drive the mix above 42 percent, consciously accepting a degradation in experience levels and readiness. Alternatively, they could drive the mix below the 20 percent RC minority status constraint, possibly compromising RC members' capacity to influence the values and perceptions of AC members.

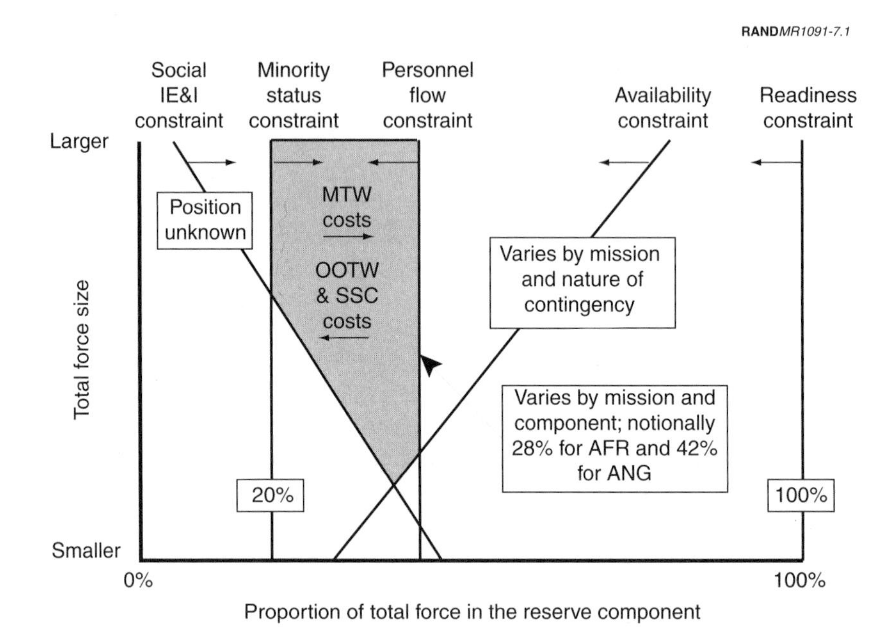

Figure 7.1—Notional Values for Force-Mix Constraints: An ANG Case

If personnel flow or availability were evaluated using different parameters, those constraints could conceivably lie to the left of the RC minority status or social IE&I constraints. There would be no feasible region. In such a case, decisionmakers would have to compromise between conflicting objectives. The most likely outcome would be to deemphasize the minority status and social IE&I constraints because the case for applying the former in the present context is less compelling and the locus for the latter is unknown.

We again stress that the specific force-mix results reported here are notional. Where possible, we used input values that we judged to be approximately correct, recognizing that we did not have the resources in this study to obtain or derive analytically rigorous inputs, especially when the inputs are likely to vary across missions. Also, because of variations across missions or MDSs, force-mix decisions cannot be made in the aggregate. They must be made for each mission or MDS individually.

Although the results reported here are notional, we believe our research provides two concrete contributions to the force-mix decision process. The first contribution is a framework for integrating the range of considerations that decisionmakers face and for gaining perspective on the arguments offered by various interest groups hoping to influence the force mix. The second contribution is a roadmap for more-detailed research into specific mission/MDS force mixes or a general model that incorporates mission/MDS-specific inputs.

Perhaps our most significant finding is that cost considerations can cut in opposite directions depending on whether the force is being optimized for major theater war preparedness or for peacetime contingency operations. In our view, peacetime contingency demands must be given more weight in force-mix decisions, especially in MDSs experiencing high deployment-related stress.

REFERENCES

Air Force Reserve Command, *Air Force Reserve Review*, 1996.

Binkin, Martin, *Who Will Fight the Next War? The Changing Face of the American Military*, The Brookings Institution, Washington, DC, 1993.

Brinkerhoff, John R., and David W. Grissmer, "The Reserve Forces in an All-Volunteer Environment," in William Bowman, Roger Little, and G. Thomas Sicilia (eds.), *The All-Volunteer Force After a Decade*, Pergamon-Brassey's, Washington, DC, 1986, pp. 206–229.

Brown, Roger Allen, William Fedorochko, Jr., and John F. Schank, *Assessing the State and Federal Missions of the National Guard*, RAND, MR-557-OSD, 1995.

Browning, James W. II, Kenneth C. Carlon, Robert L. Goldich, Neal F. Herbert, Theodore R. Mosch, Gordon R. Perkins, and Gerald W. Swartzbaugh, "The U. S. Reserve System: Attitudes, Perceptions, and Realities," in Bennie J. Wilson (ed.), *The Guard and the Reserve in the Total Force*, National Defense University Press, Washington, DC, 1985.

Butler, John Sibley, and Margaret A. Johnson, "An Overview of the Relationships Between Demographic Characteristics of Americans and Their Attitudes Towards Military Issues," *Journal of Political and Military Sociology*, 19, Winter, 1991, pp. 273–291.

Cohen, William S., *Report of the Quadrennial Defense Review*, Department of Defense, Washington, DC, May 1997.

Department of Defense (DoD), *Population Representation in the Military Services, Fiscal Year 1996*, Office of the Assistant Secretary of Defense (Force Management Policy), Washington, DC, 1997.

Department of Defense (DoD), *Total Force Policy Report to the Congress*, Department of Defense, Washington, DC, December 1990.

Department of the Air Force, "FY 99 Joint SECAF/CSAF Posture Statement," *Air Force Policy Letter Digest*, March/April 1998.

Holsti, Ole R., *A Widening Gap Between the Military and Civilian Society? Some Evidence, 1976–1996*, White Paper for the Project on U.S. Post–Cold-War Civil-Military Publications, John M. Olin Institute for Strategic Studies, Harvard University, Cambridge, Massachusetts, 1997.

Ivie, Rachel L., Cynthia Gimbel, and Glen H. Elder, Jr., "Military Experience and Attitudes in Later Life: Contextual Influences Across Forty Years," *Journal of Political and Military Sociology*, 19, Summer, 1991, pp. 101, 117.

Izraeli, Dafna N., "Sex Effects or Structural Effects? An Empirical Test of Kanter's Theory of Proportions," *Social Forces*, 62, September, 1983, pp. 153–165.

Kanter, Rosabeth Moss, *Men and Women of the Corporation*, Basic Books, New York, 1977.

Kestnbaum, Meyer, "Partisans and Patriots: National Conscription and the Reconstruction of the Modern State in France, Germany and the United States," Ph.D. dissertation, Department of Sociology, Harvard University, Cambridge, Massachusetts, 1997.

Kestnbaum, Meyer, *The Democratic Dilemma: Civil-Military Relations in the United States Since World War II*, Working Paper, Center for Research on Military Organization, University of Maryland, College Park, Maryland, 1998.

Kohn, Richard H., "The Constitution and National Security: The Intent of the Framers," in Kohn (ed.), *The United States Military Under the Constitution of the United States, 1789–1989*, New York University Press, New York, 1991, pp. 61–94.

Kohn, Richard H., *The Forgotten Fundamentals of Civilian Control of the Military in Democratic Government,* Working Paper, John M. Olin Institute for Strategic Studies, Harvard University, Cambridge, Massachusetts, 1997.

Kohn, Richard H. (ed.), *The United States Military Under the Constitution of the United States, 1789–1989,* New York University Press, New York, 1991.

Krislov, Samuel, *Representative Bureaucracy,* Prentice-Hall, Englewood Cliffs, New Jersey, 1974.

Lacy, James L., "Whither the All-Volunteer Force?" *Yale Law and Policy Review,* 5, Fall-Winter, 1986, pp. 38–72.

McDonald, Sylvia James, "Public Perception of Reserve Forces," *The Officer,* December 1996, pp. 34–35.

Moskos, Charles, and John Sibley Butler, *All That We Can Be: Black Leadership and Racial Integration the Army Way,* Basic Books, New York, 1996

Palmer, Adele R., et al., *Assessing the Structure and Mix of Future Active and Reserve Forces: Cost Estimation Methodology,* RAND, MR-134-1-OSD, 1992.

President's Commission on an All-Volunteer Armed Force, *Report of the President's Commission on an All-Volunteer Armed Force,* U.S. Government Printing Office, Washington, DC, February 1970.

RAND, *Assessing the Structure and Mix of Future Active and Reserve Forces: Final Report to the Secretary of Defense,* RAND, MR-140-1-OSD, 1992.

Reserve Forces Policy Board, *Reserve Component Programs: Fiscal Year 1996,* Office of the Secretary of Defense, Washington, DC, 1997.

Ricks, Thomas E., *Making the Corps,* Scribner's, New York, 1997.

Rostker, Bernard D., and Scott A. Harris, *Sexual Orientation and U.S. Military Personnel Policy: Options and Assessments,* RAND, MR-323-OSD, 1993.

South, Scott J., Charles M. Bonjean, William T. Markham, and Judy Corder, "Social Structure and Intergroup Interaction: Men and Women of the Federal Bureaucracy," *American Sociological Review*, 4, October 1982, pp. 587–599.

Thaler, David E., and Daniel M. Norton, *Air Force Operations Overseas in Peacetime: OPTEMPO and Force Structure Implications*, RAND, DB-237-AF, December 1997.

Wilson, James L., et al., *Considerations in a Comprehensive Total Force Cost Estimate*, Institute for Defense Analysis, Paper P-2613, Alexandria, Virginia, November 1992.

Winnefeld, James A., Preston Niblack, and Dana J. Johnson, *A League of Airmen: U.S. Air Power in the Gulf War*, RAND, MR-343-AF, 1994.

Yoder, Janice D., "Rethinking Tokenism: Looking Beyond Numbers," *Gender and Society*, 5, June 1991, pp. 178–192.